ALSO BY PHILIP L. FRADKIN

California, the Golden Coast (1973)

A River No More (1981)

Fallout (1989)

Sagebrush Country (1989)

Wanderings of an Environmental Journalist (1993)

The Seven States of California (1995)

MAGNITUDE 8

MAGNITUDE
8

PHILIP L. FRADKIN

UNIVERSITY OF CALIFORNIA PRESS

Berkeley · Los Angeles · London

University of California Press
Berkeley and Los Angeles, California

University of California Press, Ltd.
London, England

Published by arrangement with
Henry Holt and Company, Inc.

First Paperback Printing 1999

Library of Congress Cataloging-in-Publication Data
Fradkin, Philip L.
    Magnitude 8 / earthquakes and life along the San Andreas Fault /
Philip L. Fradkin.
        p.   cm.
    Previously published: New York : Henry Holt, 1998.
    Includes bibliographical references and index.
    ISBN 0-520-22119-2 (alk. paper)
    1. Earthquakes—California, Southern.   2. San Andreas Fault
(Calif.)   I. Title.   II. Title: Magnitude eight.
[QE535.2.U6F7   1999]
551.22'09794—dc21   99-18476
                                                            CIP

Frontispiece: San Andreas Fault, photograph by Philip L. Fradkin
Designed by Michelle McMillian

Printed in the United States of America

10 9 8 7 6 5 4 3 2 1

The paper used in this publication is both acid-free and totally chlorine-free
(TCF). It meets the minimum requirements of American Standard for
Information Sciences—Permanence of Paper for Printed Library Materials,
ANSI Z39.48-1984. ∞

For Carl D. Brandt

In some places Earthquake shakes the earth hard, in some he shakes it a little. For he did that in the beginning and does it now.

—Alfred L. Kroeber, *Yurok Myths*

The [tectonic] history of any one part of the earth, like the life of a soldier, consists of long periods of boredom and short periods of terror.

—Derek Ager, *The Nature of the Stratigraphic Record*

We really don't know what an 8 is going to do. But the previous experience shows us that we have been underestimating consistently all along. I think an 8 is going to be a real stunner for all of us.

——Kaye M. Shedlock, chief of the Branch of Earthquake and Landslide Hazards, United States Geological Survey, testimony before a House subcommittee following the 1994 Los Angeles earthquake

CONTENTS

PREFACE

Disclosure comes first. I am personally involved in the subject matter of this book.

For the last twenty years I have lived adjacent to the San Andreas Fault in northern California. During the previous seventeen years, when I lived elsewhere in the state, I was never far from the fissure or its many offshoots. Thus, the book is written from the perspective of a fault line resident.

It came about in the following manner.

To learn more about the natural force that evokes our deepest, most disturbing, and longest-lasting fears I set off one day on a seismic journey of discovery through time (seismic history) and space (the fault line). The result is a history of earthquakes and the science of earthquakes on one hand, and a travelogue and guide to the San Andreas Fault on the other.

I did not, however, restrict my inquiries to the single crack in the surface of the earth. Since more has happened off the fault line than on it, and all tectonic matters are related to it, I felt free to wander from that symbolic band to closely related events.

My journey was not continuous; in fact, it echoed the episodic nature of the phenomenon that periodically rattles California. Earthquakes are not consistent, so I varied directions. A single quake splits only a fraction of the long fault; I fit my travels to that portion of the fault that absorbed a specific shock. The only continuity that I could uncover was the intermittent chronology of a seismic narrative.

Additionally, I wanted to put the California story in context. That is why I began elsewhere and occasionally digressed to events that occurred out of state and in foreign lands.

There were some specific objectives.

I wanted others to be aware of the fault's physical presence and its awesome power. Another goal was to liberate earthquakes from science-speak and give them a human dimension. I wished to probe the natural catastrophe that was the dominant model for similar events, such as hurricanes, tornadoes, and floods.

To show that earthquakes were a constructive force as well as a catalyst for destruction was another objective. I was intrigued by how a culture could ignore this powerful natural agent while simultaneously being shaped by it. The reactions of people and institutions to the shaking of the surface of the earth also fascinated me.

Finally, as I got further into the subject I thought it was important to document how science had failed to deal in an effective manner with this rogue event. Unseen and unpredictable, it appeared from below without warning in our lives and left its imprint on psyches and seismographs.

My viewpoint was that of the involved layperson. I looked upon myself as a literary geologist with a notebook in one hand and a hammer in the other. I tapped away at surface formations, examined what lay underneath, and recorded the results.

Point Reyes Station, California
December 1997

MAGNITUDE 8

MAP OF THE SAN ANDREAS FAULT

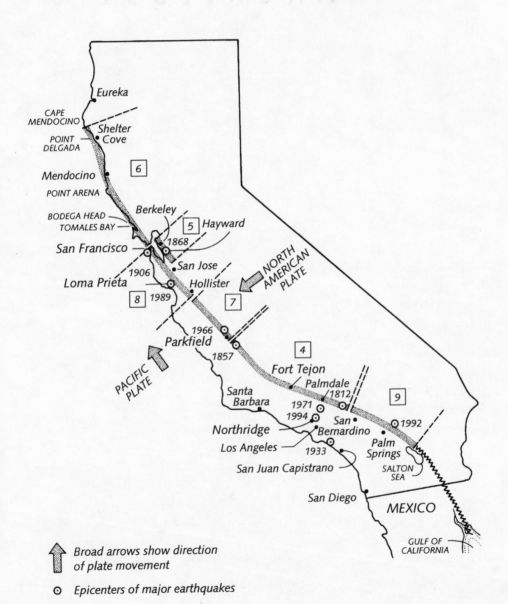

Dashed lines indicate the sections
of the fault dealt with in
Chapters 4 to 9

Eureka

CAPE
MENDOCINO

Shelter
Cove

POINT
DELGADA

6

Mendocino

POINT ARENA

BODEGA HEAD

Berkeley

TOMALES BAY

5 Hayward

1868

San Francisco

San Jose

1906

NORTH
AMERICAN
PLATE

Loma Prieta

Hollister

8 1989

7

1966

Parkfield

1857

4

PACIFIC
PLATE

Fort Tejon

Palmdale

1812

Santa
Barbara

9

1971

1994

Northridge

San
Bernardino

1992

Los Angeles

Palm
Springs

1933

San Juan Capistrano

SALTON
SEA

San Diego

MEXICO

GULF OF
CALIFORNIA

Broad arrows show direction
of plate movement

⊙ Epicenters of major earthquakes

SAN FRANCISCO 1
(the Present)

THE DAY CALIFORNIA RAN OUT OF LUCK

It was an ordinary October weekday late in the dry season, hot and windy and fogless as it can be at that time of year. Earthquake weather, some would later say, but that was hindsight.

The time was 2:14 P.M. Small children played in elementary-school yards, older students studied in high schools or universities, and adults labored in downtown high-rises. The rush-hour traffic was just beginning to build on the Bay Area's numerous bridges and elevated freeways.

A technical writer stood in a window of an office building overlooking Market Street. She stared uncomprehendingly at a small wave that moved down the broad commercial corridor, lifting and then depositing everything in its path. Nothing was disturbed. A moment later the destructive wave hit.

An emergency planner for a large corporation ascended from the Bay Area Rapid Transit station. He heard a deep rumbling noise that seemed to come from underneath the city. Almost instantaneously the first shock struck him at the top of the stairs. He clutched the handrail and looked across Market Street. Like a vertical snake making its way across the azure sky, there were three bows in the undulating Crown Zellerbach Building.

Earthquake, he thought. He ran, then crawled through a shower of concrete and glass to the doorway of a rapidly disintegrating forty-story building whose steel welds had been cracked in the 1989 quake.

For the two German couples in the rental car, there was no warning in the small coastal village of Point Reyes Station thirty miles to the north of the city. The car windows were closed. The air conditioner and radio were on, although the light rock music had just turned to static. The car's shocks absorbed the first minor blows.

The driver parked across the street from a restaurant in the shade of an abandoned two-story brick building. The heavy Spanish roof tiles and red bricks buried the four tourists in an avalanche of debris loosened by a force that was the equivalent of a large underground nuclear explosion.

Twenty miles to the south of San Francisco the afternoon wind ruffled the surface of Lower Crystal Springs Reservoir. A keen observer would have noticed the crosshatched pattern of the waves that overlay the fault line. The fish jumped clear of the water's disturbed surface, as if their natural habitat had become a hot frying pan.

As the ground began to sway, the suburban strollers and joggers on Sawyer Camp Trail, which parallels the eastern shoreline of the reservoir, felt nauseous and staggered. They were knocked to the ground when the earth shook with a motion akin to riding an out-of-control subway train without the benefit of support. They curled protectively into fetuslike shapes on the vibrating asphalt pavement of the trail. Some lay as they were; others attempted unsuccessfully to rise as the earth shuddered violently for one minute and twenty-three seconds.

The worst was yet to come for these huddled figures.

The powerful earthquake was centered under the San Andreas Reservoir, just upstream from Lower Crystal Springs Reservoir. The long fault derived its name from the San Andreas Valley that had lain submerged for 130 years beneath the reservoirs.

Most of the casualties occurred shortly after the earthquake, when two dams collapsed and fires erupted throughout the region.

The San Andreas Dam, completed in 1869, barely survived the 1906 San Francisco earthquake. It was no match for the magnitude 8 event whose shallow source lay underneath the old earth-filled structure.

The west side of the dam was jerked twelve feet northward. The face crumpled. From slow motion to fast forward and accompanied by a roaring sound resembling multiple jet engines, the water cascaded downstream toward Crystal Springs Dam, engulfing the stunned suburbanites on the trail.

The second dam, a concrete-block structure, was built in 1890 and had been designated a California Historic Civil Engineering Landmark. Crystal Springs Dam lay just three miles west of downtown San Mateo. Earthquakes in 1906 and 1989 did not budge the 154-foot-high structure. The old waterworks, however, needed an extensive overhaul in a time of declining budgets; none had been forthcoming.

The earthquake opened existing cracks in the old dam and the outlet towers and dislodged the rickety bridge that spanned the dam. It fell on top of the structure, blocking the spillway. With the rapid inundation of the floodwaters from San Andreas Dam, the tottering structure collapsed.

The massive tragedy that unfolded below the dam fulfilled a water official's prophecy that, should such a disaster occur, "We would be retrieving San Mateo, Hillsborough, and Foster City from San Francisco Bay."

The initial discharge approaching one million cubic feet per second generated enough erosive force to reduce all human and natural artifacts to bare earth. A wall of water cresting at a height of 112 feet set up an additional vibration in the earth as it raced down the narrow canyon carrying house-sized chunks of concrete and whatever flotsam it could claim along the way. The cascading wave lessened in speed and size as it flared out onto the crowded San Mateo County shoreline.

While crushing structures and overwhelming people in its froth, the surging water also extinguished the many fires ignited by the earthquake.

The seismic shaking had broken natural-gas mains and severed connections to individual homes. Petroleum products leaked from tank farms and either ignited or flowed into the bay, where they were joined by other toxic materials and raw sewage that poisoned wildlife. From the Inverness Ridge to the north, Mount Diablo in the east, and the Santa Cruz Mountains to the south, wildfires fanned by strong afternoon winds and feeding on the dry vegetation escalated rapidly, igniting rural propane tanks along the way.

The sun was a sickly red disk behind the roiling clouds of dust, smoke, and sulfurous fumes. The sky darkened early that afternoon. There was no electricity in the entire Bay Area to ease the blackness—just the flames that danced like guttering candles in the rolling terrain. The flickering landscape popped and crackled.

Within the city the largest fire spread rapidly eastward from the outer Sunset District near the ocean. No firewalls separated the abutting wood-frame homes that stood in units of twenty-four with a combined floor space of over thirty thousand square feet. A former San Francisco fire chief, estimating damage in a future quake, had said: "Under major fire conditions this fire load [meaning one unit] would tax the best efforts of an entire metropolitan fire department. There are, unfortunately, thousands of such houses across the city."

They fell like flaming dominoes. There were no operable phones to report the conflagration that soon coalesced into a raging firestorm, feeding upon itself and consuming everything in its downwind path toward the stilled commercial heart of the city. The red fire-alarm boxes on street corners were useless because of damage to underground conduits.

Had engine companies been sent—assuming there was an operable dispatch center—they could not have gotten far. Streets were littered with debris and glass that hindered passage and cut tires. There would have been little or no water once they arrived. Not only had the reservoirs been emptied, but also distribution lines within the city were severed, as was the Hetch Hetchy Aqueduct, which carried water from the Sierra Nevada to many Bay Area communities.

Fires, beyond any human's ability to control, raged throughout the city and the hinterlands. All that could be done was to let them burn and pray for diminishing winds and the return of the fog.

Other than on foot, it was impossible to flee or enter the city. Freeways and side streets were buckled, overpasses were down, landslides covered roadways, and either the spans of all five bay bridges or their approaches had snapped or sunk into the jellied soil. Motorists panicked. There were numerous accidents at intersections where no traffic lights functioned.

Bay Area airports, located on susceptible foundations, were a shambles. The answer for the frantic media, it appeared, was to land in Sacramento or Los Angeles, strap mountain bikes onto rented sports utility vehicles, and then descend into this blazing, waterless hell that had once been a functioning city beloved by all.

The consultant and planner, the German tourists, and the suburbanites perished along with 28,000 others in northern California. It was the greatest domestic tragedy in this country's history, excluding the Civil War. The magnitude 8 earthquake was approximately the same size as the 1906 event. The difference was that at the start of the century there were 660,000 people living in the Bay Area, and now there were more than 6,000,000. Some of the structures had been made more earthquake-resistant; others hadn't.

As darkness fell that first night, there was an eerie human silence and the sound of crackling flames. A French newspaper declared that San Francisco had ceased to exist, but that was not the case. Once again, with much fanfare and denial, it would be rebuilt.

Is this fiction or fact?

All incidents and suppositions are either extrapolations from smaller earthquakes or scenarios concocted by federal and state authorities for a 1906-type event. The quotes come from documents and people. I have simply blended them into a reasonable, if somewhat horrific, narrative account.

Quantification of such a disaster is a crapshoot. The variables are immense and many are unknown. Twenty-eight thousand deaths is a conservative figure. A 1981 study estimated 22,000 to 33,000 deaths from the single dam I use in my example. Population has increased since then, and additional dams could fail. The estimates of deaths, not counting those from dam failures, have ranged upward to 12,000. Conceivably, the death toll on paper could be 45,000.

The approximate ratio of deaths to injuries is three hospitalized and thirty nonhospital injuries per death. Given my estimate, this would mean 90,000 serious injuries. Factor in off-duty personnel who could not get to hospitals and medical facilities that would be destroyed or badly damaged, and it seems likely that many would suffer. The slightly injured and homeless might run to over a half million people.

The most recent damage estimate for such a quake, again not taking a dam failure into account, ranges from $170 billion to $225 billion. The present record for earthquake damages in this country is $20 billion to $40 billion from the more moderate 1994 Los Angeles quake, which purportedly emptied insurance company coffers. Obviously there would be considerable national and international impact, considering damage to Silicon Valley and trade with Pacific Rim countries alone.

I realize this account is difficult to accept, but history demonstrates that it will occur in a similar form and place in the near future. No one can know with any degree of certainty when or where such a catastrophe will strike.

TERRA NON FIRMA 2

EARTHQUAKE COUNTRY

Description, definition, and mythological concepts begin this probe into the danger of place.

Among the rift zones of the world, the San Andreas Fault is quite unusual. Its distinctiveness is derived from its physical continuity. For hundreds of miles that giant rut—seemingly made by a single carriage wheel that varied only slightly from its course, becoming fainter or clearer depending upon the composition of the soil and rock—is recognizable as a feature of the landscape. Not a single place, but rather the accumulation of many related features bespeaks its awesome power.

The U.S. Geological Survey (USGS), the federal agency within the Department of the Interior that deals with seismic matters, refers to the fault system as the best-known and most studied plate-tectonic boundary in the world. The San Andreas is the longest and most active of the numerous faults in California that collectively form the border between the Pacific and North American plates.

There are three components to that band. From widest to narrowest, there are the fault system, fault zone, and fault. All the immediate offshore and onshore faults along the tectonic boundary comprise the San Andreas Fault system, which is fifty miles wide at the latitude of San Francisco. The fault or rift zone varies from one-third to a half mile, that being the area of highly sheared rock. The fault, fault line, fault trace, ot fault branch is the most recent discernible break in the crust of the earth. A visible fault can be a single or a double furrow or a series of parallel fissures resembling half a chevron. Many faults, termed *blind*, do not break the surface.

Mileage helps define the compelling presence of the San Andreas. While the fault system runs from Cape Mendocino in the north to the Colorado River delta in Mexico, where it drops into the Gulf of California, the single strand of the fault begins just north of Point Delgada and stops short of the border at the Salton Sea. The fault system is slightly longer than 800 miles, while the fault is 660 miles in length.

Celebrated in films, novels, and journalism, the fault has achieved a notoriety of legendary proportions. It is also a symbolic presence. Like palm trees, it defines California, particularly for outsiders.

When a poll conducted by Fodor's, the guidebook publisher, designated New York City as the runner-up in the category of places travelers least liked to visit and California as the second most desirable place, *The New York Times*, a booster of the former and denigrator of the latter, editorialized: "The lofty towers and the shopping and the Statue of Liberty are still thrilling. Of even deeper significance, New York City does not lie along the San Andreas fault, unlike the second favorite destination."

Beyond metaphor, the fault possesses a tangible grandeur that is a product of its conspicuous visibility over a great distance. From the latitude of a rain forest environment to the low desert, the fault bisects surprisingly few urban areas. Rising from the ocean in Humboldt County, piercing the mile-high pine forests in northern Los Angeles County, and then dropping to below sea level in the sands of Imperial County, the fault resembles a long scar incised along the soft underbelly of California.

What is particularly noticeable about the fault is the linearity of topographic features. Valleys, bays, lakes and ponds, notches in ridgelines, and the abrupt ends of ridges and mountains are arranged in a straight line or a gently curving arc to form an echo of the coastline. No other single landscape feature, except the shoreline, is as continuous in California.

There are benefits from tectonic movement, among them being the variety of landscape forms that distinguishes California from other terrains. Without the San Andreas and its many associated faults, California's landscape would be as dull as Iowa's. There is another advantage. Earthquakes tend to discourage outsiders from migrating to California and they encourage the fainthearted to flee, thus contributing to the depopulation of the state. The rift-zone valleys provide a flat place to build houses, schools, universities, hospitals, churches, roads, freeways, aqueducts, and dams. What the fault gives, however, it can also take away.

Besides their physical features, earthquake faults have been the focus of much conjecture over the centuries. Various theories have been advanced to explain why the surface of the earth moves. One replaced another with regularity. Plate tectonics, a theory that emphasizes the dynamic aspects of the continents and oceans coming together and parting, is the current orthodoxy.

This theory holds that there are seven large plates and a number of smaller ones that pave the surface of the earth. These slabs are constantly moving relative to one other. Thus, most earthquake and volcanic activity occurs at plate boundaries, such as along the San Andreas Fault and to the north in the volcanic Cascade Range.

There are three types of movement: the plates pull apart, they move horizontally past each other, or one dives under another—a process called *subduction*. Only the first of these types of movements does not occur in California. The activity is mostly horizontal along the San Andreas Fault system. From Cape Mendocino north, the Pacific Plate is diving under the North American Plate in a region known as the Mendocino Triple Junction.

What does it feel like to live near, adjacent to, or on top of such a powerful force? Most people are unaware of its existence, since there is an almost total absence of signage, and damaging events are infrequent. Others give very little thought to its presence. These were the prevalent attitudes that I encountered along the fault line, where the transient nature of the populace does not lend itself to long-term memories.

There were some exceptions. Psychological aftershocks were expressed in different ways: guilt (a mother thought of herself before her baby in a quake); fear (a man frequently awakens trembling in the night); prudence (a professional couple moved off the fault when they had a child); fantasy (a young girl who heard her grandfather's school was destroyed in the 1906 quake wished the same would happen to her school); helplessness (a woman cried for a week because her insurance company limited the earthquake coverage on her home, thus endangering her nest egg); and excitement (two persons I encountered said they were turned on by earthquakes).

THE NATURE OF SEISMIC EVENTS

Earthquakes are prima facie a natural disaster. Yet the term "natural" in this context is a misnomer. Earthquakes, volcanoes, wildfires, tsunamis, landslides, avalanches, floods, droughts, blizzards, hurricanes, tornadoes, cyclones, and lightning are disasters or catastrophes only when humans and their works get in the way. In reality, they are natural processes.*

Earthquakes are the most all-inclusive of such events. Frequently, earthquakes accompany active volcanoes. Flames erupt and fires spread when utility lines are ruptured. Tsunamis are spawned by the violent movement of the earth and cause flooding, as do dams that are breached by quakes. The dust rises, the land subsides, and the snow descends when the earth shakes.

Thus, a study of earthquakes is very close to being an investigation into all forms of natural catastrophes. In fact, this archetype of sudden-impact hazards is frequently used as a model in the newly emerging field of disaster studies.

Like an iceberg, the vast bulk of this phenomenon is hidden from sight. Earthquakes are the result of tremendous forces deep within the earth that are invisible to the naked eye and only dimly understood by the human intellect. What moves has become clearer in the twentieth century; why and when remain a mystery.

The fear of the unknown is the greatest fear. Temblors strike without warning or at most, without a reliable warning that we can discern. Our senses and their technological extensions are no defense. Comfortable in our familiar surroundings, we are blind to their shazamlike appearances. These periodic upheavals maim, kill, and destroy within moments; yet they haven't significantly altered the course of modern history. Ancient times, when civilizations were more tightly clustered, were a different matter.

Some scientists and scholars believe that earthquakes or their accompanying tsunamis caused the Red Sea to part for Moses and the fleeing

*Tsunami, the Japanese word for large waves generated by an earthquake or volcano, rather than tidal wave, is the proper term since no tidal action is involved. With the above caveat in mind, I shall continue to use the terms natural disaster, etc., because they are widely employed.

Jews, breached the walls of Jericho and Troy, and destroyed Sodom and Gomorrah. Armageddon and the Apocalypse were associated with earthquakes. An earthquake struck when Christ died on the cross. The ancient fortress city of Megiddo, located at a strategic position on a branch of the Dead Sea Fault, was destroyed and rebuilt numerous times after such violent disruptions as seismic events.

The Minoan civilization on the Aegean island of Thíra—possibly Plato's Atlantis—was torn apart by earthquakes and then vaporized by volcanic eruptions. The resulting tsunami and ashfall also wiped out the Minoans on nearby Crete. Their civilization was absorbed by the Greeks. The scientist-writer team of Charles Officer and Jake Page wrote in *Tales of the Earth*: "No earthquake that we know of has changed the course of civilization, as the volcanic eruption at Thíra evidently did, but it is safe to say that no geologic phenomenon has taken a greater toll of human lives than earthquakes."

Earthquake casualties predominate in third world countries, while property damage is highest in developed nations.

Although perhaps 60,000 died in the great Lisbon earthquake of 1755, Europe suffered more from wars, epidemics, and famines during the middle years of the current millennium. The seismic toll has been greater elsewhere. For the four centuries following A.D. 1400, there were 110,000 earthquake deaths in Europe compared to 1,200,000 in China.

No country has suffered greater seismic losses than China: 13,000,000 deaths over the three thousand years that records have been kept; 830,000 killed in one temblor in 1556; and upwards of 250,000 casualties in another in 1976. In the two hundred years of recorded California history, there have been fewer than 6,000 deaths.

When Americans shake their heads in disbelief at California's fragility without considering the earthquake history of the remainder of the world, they are being geocentric. The chance of perishing in an earthquake in California is 1 in 600,000, while the odds of dying in an automobile accident are 1 in 20,000. Still, California is considered as the riskiest place to live in this country because of an extensive array of natural and human hazards, the first being the state's "infamous seismic zones."

The large number of catastrophes in California should be no surprise, both to those living within the state and outsiders.

California contains every type of landscape province and its accompanying dangers: blinding sandstorms and flash floods in the desert, blizzards and avalanches in the High Sierra, volcanic mountains in the Cascade Range, the Coast Range riven by fault lines, and large rivers and smaller creeks capable of flooding in flat valleys.

With a wet season and a dry season making up a given year, excesses of precipitation and drought bring floods and fires. Across the huge expanse of open water stretching all the way to Asia come El Niño, hurricanes, and other unstable masses of air bearing such minor disturbances as lightning storms, hail, and tornadoes.

Combine all these natural forces with the largest population in the nation and the result can be nothing else but periodic chaos, a fact that is not widely advertised.

EARLY HISTORY AND MYTHOLOGY

Earthquakes were first associated with animals. The world was depicted as resting on some creature's back. The animal moved; the earth shook. Depending on the mythology of the particular culture, this creature could be a tortoise, buffalo, hog, frog, fish, crab, mole, or serpent.

The results, of course, were catastrophic; they could also be beneficial. By shifting position the god or hero-animal got a better purchase on the earth, thereby assuring it continued stability.

In the earthquake-prone Mediterranean region, the fourth-century B.C. world of the Greek philosopher Aristotle consisted of four elements: earth, air, fire, and water. His theory of earthquakes was constrained by this short list. Wind, fanned by subterranean fires or the heat of the sun, was either exhaled from or forced into cracks in the earth, thus causing the earth to shake. Aristotle's hypothesis, contained in his *Meteorologica*, dominated thinking on the subject up to the middle of the last century.

An offshoot of Aristotle's theory maintained that the collapse of underground caverns caused quakes when the supports that held up the earth were consumed by fire. Yawning chasms that could swallow large objects appeared momentarily, then snapped shut. The fear of such abysses persists to the present, despite lack of hard proof of their existence.

In the New World, the Indians of California experienced the most earthquakes. For such an active seismic zone there are few surviving Native

American accounts, which might be a comment on anthropologists not asking the right questions.

Thunder and Earthquake were two characters in the myths of two widely separated tribes, the Yokuts of the southern San Joaquin Valley and the Yuroks of the Northwest Coast. The land shook when Earthquake ran while at the same time Thunder accompanied him with various pyrotechnics in the sky.*

As the tale was recorded by University of California anthropologist Alfred L. Kroeber, the Yuroks saw the interaction between Earthquake and Thunder as a playful competition with serious overtones:

> Then Earthquake thought: "How will it be about the earth?"
>
> Thunder came and said, "It will be best if I help you when you shake."
>
> Earthquake thought: "Perhaps it will not amount to anything if he helps me."
>
> Thunder said, "It will be well, for I shall be running all over the world, and it will be good like that."
>
> Earthquake said, "Well, I shall tear up the earth."
>
> Thunder said, "That's why I say we will be companions, because I shall go over the whole world and scare them. Wherever I know people live, I shall go, upstream or across the ocean, for I brought something to be seen at night; at Pulekuk I brought it."
>
> Earthquake said, "If I see the earth tilt, I can level it again. That is what I shall want to do."
>
> Thunder said, "I will begin to run. Listen." So he began to run, and he listened. It seemed as if the sky began to fall, so hard did Thunder run, and leaped on trees and broke them down.
>
> Earthquake stayed still to listen to his running. Then he said to him, "Now you listen: I shall begin to run." He started. He shook the ground. He tore it and broke it to pieces, because he did not wish us to be about.

The Yuroks, who lived within the zone of subduction, had their own way of understanding how and for what purpose the landscape was altered. The prairie near the ocean subsided when Earthquake stepped on it, and the

---

*The sound of distant thunder and flashes of light accompany some earthquakes.

result was more productive terrain, that being the lagoons along the North Coast. This could have been a case of the land subsiding during a huge earthquake, a tsunami inundating the coast, or both occurring at the same time. Scientists now think that a giant temblor or a series of smaller quakes shook the Northwest between 1700 and 1730, causing parts of the coast to subside.

California Indian tribes interpreted the 1906 earthquake differently than the dominant white culture. To the Yurok the temblor was punishment for stealing many of the tribe's artifacts and putting them on display in museums in Berkeley and San Francisco. The Washo in the Sierra Nevada saw it as payback for theft of natural resources. To the Wintun of the Sacramento Valley, the quake was either the stretching of the earth to accommodate the rapidly multiplying whites, or the beginning of the obliteration of all races.

THE OLD AND NEW WORLDS 3
(1580–1812)

LONDON

The word *earthquake* began to make its appearance in the English language in the fourteenth century, when it was associated with pestilence at the time of the Black Death. It was from the English, who were periodically plagued by earthquakes, that our nascent culture inherited its seismic beliefs.

The Easter earthquake of 1580 caused a London rector to write a commemoration about preparedness metaphorically entitled "A Bright Burning Beacon, forewarning all Wife Virgins to trim their lamps against the coming of the Bridegroom." The poet John Milton and the novelist Daniel Defoe, author of *Robinson Crusoe*, mentioned earthquakes in their texts.

Shakespeare was sixteen at the time of the 1580 tremor, and references to earthquakes show up in a number of his works. The bard was influenced by Aristotle's concept of hot winds, which may have been the genesis of the concept of earthquake weather. In *Henry IV*, Hotspur declares:

> *Diseased nature oftentimes breaks forth*
> *In Strange eruptions; oft the teeming earth*
> *Is with a kind of colic pinch'd and vex'd*
> *By the imprisoning of unruly wind*

> *Within her womb; which, for enlargement striving,*
> *Shakes the old beldam earth, and topples down*
> *Steeples and moss-grown towers.*

What seems like a topical reference in *Romeo and Juliet* is used to date the play. Juliet's nurse recalls an unforgettable day and foreshadows the coming tragedy in act 1:

> *'Tis since the earthquake now eleven years;*
> *And she was weaned (I shall never forget it),*
> *Of all the days of the year, upon that day.*

Scholars had recorded quakes in 1580, 1583, and 1585 from which to choose. Using other clues in the text, they selected the 1585 Kent temblor. By adding eleven years, they wound up with a 1596 date for the play.

A pamphlet, titled "Perpetual and Naturall Prognostications of the Change of Weather," was translated from the Italian. (Italy being quite earthquake prone) and printed in London in 1591. It was a sort of sixteenth-century version of the *Farmer's Almanac* and claimed to draw upon the accumulated wisdom of Cato, Aristotle, Virgil, Plato, Agrippa, and "many others."

An earthquake was likely when a number of signs "do concur together in one." The portents included "troubled and thick" water in wells; the swelling of the sea "with billows and great waves as in a tempest"; red and fiery clouds; a "dark, red, bloody, or fiery" sun or moon; and "any terrible noise heard in the air like unto howling bellowings." Beware "if the weather be very quiet, still, and calme for a month or two together without any motion or stirring of winds," and also watch for the unusual movements of "all birds and forefooted beasts."

The year 1750 was known as "the year of earthquakes" in England, a designation that would later be attached to the year 1812 in this country. There were five noticeable shocks, of which two were in London. From the date of the first quake on February 19 through the end of the year, nearly fifty articles and letters on earthquakes were read before the Royal Society. Horace Walpole, a man of letters, noted in a personal communication on April 2: "Several women have made earthquake gowns, that is, warm gowns to sit out of doors all to-night." (Outdoors was considered a safer refuge than indoors.)

LISBON

Following the year of earthquakes, there was a five-year interlude of seismic quiescence in the Old World. Then a massive disturbance shook the Continent. The Lisbon earthquake of 1755 severely tested the pervading optimism and theologies of Europe. It was the Age of Reason and the Enlightenment. Europeans were shaken to their collective core by the unexpected cataclysm that seemingly defied reason.

The medieval city of Lisbon was a center of Catholicism, and a staggering amount of material wealth was stored in its magnificent churches, palaces, and warehouses. "Wealth, the Inquisition, and the worship of images: to an appreciably large section of the outside world Lisbon was famous for these three things," wrote T. D. Kendrick of the British Museum in his account of the earthquake and its theology.

November 1, 1755, was All Saints' Day, and much of the city's population was gathered in the many churches for mass. Shortly before 10 A.M. a sound resembling distant thunder was heard. It was followed by a series of tremendous shocks that lasted for an unheard-of six or seven minutes. Other convulsions occurred throughout the day, which had dawned clear and bright. A dark cloud quickly settled over the city.

Those who weren't immediately crushed fled outdoors, some to the newly built granite quay beside the Tagus River, which emptied into the Atlantic. A series of tsunamis rolled in from the ocean, uprooted the many anchored ships in the harbor, and swept the quay and other low-lying areas in the city. Fire, the third destructive force, erupted when candles ignited curtains and fallen wooden beams. The city burned for nearly one week. One-fourth of the population of 240,000 perished.

Kendrick described a scene that could have been the result of a magnitude 8 earthquake:

> It was as savage a gutting of the heart of a city as can be found anywhere in the previous history of Europe, and after this ferocious blaze had done its work the richest and most thickly settled populated district

of Lisbon was a charred desert of smoking ruins with the dead bodies of
[thousands] of the inhabitants lying beneath the ashes and cinders of
their homes.

The effects of the earthquake were experienced on three continents.
There were deaths in North Africa. The level of the ocean rose and fell in the
West Indies. There was severe damage in neighboring Spain; and the temblor was felt in France, Switzerland, and Italy. Inland waters were disturbed
as far north as Finland.

The Portuguese had their own special forms of denial that they practiced
after the temblor. A Jesuit priest who had challenged the authorities was
accused of blasphemy. He was strangled and roasted in the public square
during a torchlit auto-da-fé, the burning of heretics at the stake being a specialty of the Inquisition. A blasphemous book claiming that God's grace had
been withdrawn from Lisbon and an effigy of its Portuguese author, who had
escaped to England, were thrown into the same bonfire.

Lisbon and California served as the Sodoms and Gomorrahs for their
times. From outside Portugal, the charges of pride and insouciance, which in
time would also come to be leveled at the inhabitants of San Francisco and
Los Angeles, were hurled at the people of Lisbon. They shouldn't have been
living in such a "charming, flower-bedecked place," was the thinking.

The earthquake was the defining event for a remarkable generation of
thinkers. Afterward, the outstanding intellects of the century engaged in the
greatest theological debate ever waged over issues raised by a natural catastrophe. The musings of the philosophers were the most far-reaching legacy
of the Lisbon earthquake, for the question of who was to blame—an
almighty being or frail humans—is asked after every such disaster.

The plaintive cry arose from the ashes of Lisbon: "Wherefore hath the
Lord done this?"

The acidic Voltaire was first to take up his pen. Within weeks of the
earthquake, he attacked the prevailing *tout est bien* philosophy in a poem.
The popular philosophy in 1755 that all was for the best because a new city
shall be built upon the ashes was also the rallying cry following the 1906 San
Francisco earthquake. Disregard the silly cliché, *tout est bien*; regard the

truth, *"Le mal est sur terre"* wrote Voltaire in a comment that had a more timeless relevance.

Concerning Voltaire's poem on the Lisbon earthquake, Kendrick wrote that it "was widely read and discussed, for it was the comment of one of the wisest men in Europe on a disaster that had shocked western civilization more than any other event since the fall of Rome in the fifth century."

To Rousseau, a defender of the state-of-bliss school, Nature was not to blame. It was the fault of the inhabitants of Lisbon, who chose to live in crowded, multistoried, earthquake-prone apartments rather than smaller houses in the safer countryside. Rousseau, the philosopher of divine Nature, argued that God was a part of Nature, but He was not concerned with the individual's fate. God's proper domain was the supervision of Providence. In this manner, Nature and God were absolved of any blame for the earthquake.

Ridicule was the weapon of choice the tart-tongued Voltaire employed against Rousseau and like-minded philosophes in his satire, *Candide*. Pangloss, a naively optimistic philosopher, and Candide, the innocent youth, are shipwrecked on the outskirts of Lisbon. They feel the earth tremble and watch as the sea rises "in foaming masses" and the "whirlwinds of flame and ashes" devour the city. They wander among the smoking ruins and aid the inhabitants. Then:

> Some citizens they had assisted gave them as good a dinner as could be expected in such a disaster; true, it was a dreary meal; the hosts watered their bread with their tears, but Pangloss consoled them by assuring them that things could not be otherwise. "For," said he, "all this is for the best: for, if there is a volcano at Lisbon, it cannot be anywhere else; for it is impossible that things should not be where they are; for all is well."

Kant and Goethe chimed in from Germany.

A young Kant wrote three short papers on the subject. Still an adherent of the philosophy of optimism, a position that would change, the philosopher-scientist pointed out that earthquakes were a part of nature and must be accommodated. There were benefits. The subterranean fires that caused the ground to tremble also provided the hot springs and baths that cured. Moreover, all humans must die, and property is not everlasting. Kant noted certain earthquake phenomena. Animals were visibly affected before a quake: birds, rats, mice, and worms fled their accustomed habitats.

The poet Goethe was six years old at the time of the Lisbon quake. Through the viewpoint of that child, he later described the lingering terror and the theological confusion that follow such events:

> Perhaps never before has the Demon of Fright so quickly and so powerfully spread horror throughout the land. The little boy, who heard everybody talking about the event, was deeply impressed. God, the creator and preserver of heaven and earth, God, said to be omniscient and merciful, had shown himself to be a very poor sort of father, for he had struck down equally the just and the unjust. In vain the young mind sought to combat this idea; but it was clear that even learned theologians could not agree about the way in which to account for such a disaster.

NEW ENGLAND

The seismic record for the New World commenced in 1627, seven years after the Pilgrims landed in what became known as Massachusetts. A sharp quake was noted twenty-five miles northeast of Boston in the direction of Cape Ann. Of Native American accounts, or the lack thereof, a member of the Boston Society of Natural History said in 1869: "Indian tradition gives us nothing of any value, for people in an uncivilized state do not note the paroxysmal changes of Nature, though the weather signs are often carefully observed."

On June 11, 1638, a shock of about the same magnitude was first heard, then felt by an expanding Anglo population throughout New England. It was centered in Canada's St. Lawrence River valley. The sound resembled "a rumbling noyse, or low murmure like unto remoate thunder."

Brick chimneys, particularly vulnerable because of their protuberant shape and lack of reinforcement, were knocked askew; and dishes were broken throughout the region. People held on to trees and posts to keep from falling. The aftershocks continued for twenty days.

On December 19, 1737, a quake toppled chimneys and rang church bells in New York City.

As the population grew and became more urban, the damage increased.

On November 18, 1755, eighteen days after the Lisbon quake and before the news of the tragedy had reached the Americas, another earthquake centered off Cape Ann was felt from Nova Scotia to Maryland. It was preceded by a noise resembling distant thunder. Chimneys tumbled in Boston, and over one thousand roofs were damaged. Stone fences were dislodged in the countryside, particularly on a line from Boston to Montreal. A tsunami swept south to the West Indies, where vessels and fish were stranded on land after its retreat. The shock was so great onboard a sailing ship two hundred miles off the cape that all aboard thought they had run aground in open water.

The quake was most noticeable in Boston. John Winthrop, professor of mathematics and natural philosophy at Harvard University, said, "The bed on which I lay was now tossed from side to side; the whole house was prodigiously agitated; the windows rattled, the beams cracked, as if all would presently be shaken to pieces."

For residents of Boston and students at Harvard, Winthrop gave a lengthy lecture on the cause of the earthquake one week later in the university chapel. Winthrop's primary interest was astronomy, but he was the closest that colonial America could come to mustering an authority on earthquakes. It was his job to allay fears, purge superstitions, and assure the populace that all was well.

Aristotle's ideas, transported to the colonies via England, formed the basis for the Harvard professor's causation theory.

Underground volcanic fires heated water that was converted into steam. What followed, said the professor, was the production of vapors that "will rush along through the subterraneous grottos, as they are able to find or force for themselves a passage; and by heaving up the earth that lies over them, will make that kind of progressive swell or *undulation,* in which we have supposed earthquakes commonly to consist; and will at length burst the caverns with a great shaking of the earth."

As with the anal expulsion of air by the animal population, Winthrop said, such "explosions" of vapor were healthy for the planet since they "open its pores." Echoing the *tout est bien* school on the Continent, Winthrop saw a confluence "between God's government of the *natural* and of the *moral* world."

Church attendance soared and increased again after news of the Lisbon quake arrived in late December. The governor of Massachusetts declared "a day of humiliation and prayer, in acknowledgment of the distinguishing mercy of God, and in submission to his righteous judgments." A minister at South Church thundered, "O! there is no getting out of the mighty Hand of God! If we think to avoid it in the Air, we cannot in the Earth."

NEW MADRID

The earthquakes on the East Coast were polite sneezes compared to the massive convulsions that rocked the lightly populated central Mississippi River valley in late 1811 and early 1812. The New Madrid, Missouri, earthquakes—felt throughout the Midwest and the East—were a distinct warning that all sections of the nation were susceptible to great seismic disturbances.

The magnitude 8 earthquakes were centered around the hamlet of New Madrid (pronounced MAD-rid). They were stronger and were felt at far greater distances over a longer period of time than any other temblors in the recorded history of the contiguous states.*

The three major shocks were followed by two hundred damaging aftershocks. There have been more than twenty damaging quakes since that time in what would be assumed—according to plate tectonics theory—to be the stable heart of the continent. A rift zone of local proportions has recently been detected buried underneath the thick alluvial deposits of the Mississippi River.

The first earthquake shook the eastern half of the continent shortly after 2 A.M. on December 16, 1811; the second took place on January 23, 1812, at 9 A.M.; and the third, possibly the greatest, was recorded at 3:45 A.M. on February 7, 1812. The epicenter for the first quake was in northeastern Arkansas while the last two were centered in Missouri's bootheel.

---

* See the discussion of magnitude and the magnitude table in the appendix. The magnitudes were later derived from physical descriptions of the damage, since there were no precise seismographic instruments at the time.

The shocks appeared in historical accounts in twenty-seven states. A Michigan judge later journeyed into northern Canada and wrote·that residents there had experienced nine shocks, the same number he had tallied in his home state. The earthquakes were felt in Boston, New York ("A smart shock," noted the *New York Post* of the February 7 quake), Charleston (the South Carolina city would be flattened by a midplate quake in 1886), and New Orleans, to name just a few places.

Flashes of light were observed in Savannah, Georgia, and low, rumbling noises were heard as far away as Washington, D.C. Recorded observations were absent from west of the Mississippi, there being few literate English-speaking inhabitants and travelers in that direction at the time.

Politics and politicians were engaged by these earthquakes.

The Shawnee chief Tecumseh, who advocated Native American unity in the face of white encroachments, used the earthquakes for his political purposes. He preached throughout the region that the Great Spirit, who speaks in thunder and swallows villages, was angry with their enemies.

In the nation's capital, one thousand miles to the east of the epicenters, furnishings were shaken and inhabitants awakened. The scaffolding erected around the Capitol collapsed. President James Madison wrote to his friend Thomas Jefferson a few hours after the February 7 shock:

> The reiteration of Earthquakes continues to be reported from various quarters. They have slightly reached the State of N.Y. and been severely felt W. and S. Westwardly. There was one here this morning at 5 or 6 minutes after 4 o'C. It was rather stronger than any preceding one, and lasted several minutes; with sensible tho' very slight·repetitions throughout the succeeding hour.

The year 1811 became known as the annus mirabilis, and 1812 was called "the year of the earthquakes," not only because of events in the Midwest but also because of deadly temblors in California and South America.

A comet, the harbinger of calamity, flashed across the fall and winter sky, and there was an almost total eclispe of the sun. (A comet and eclipse of the moon were followed by a mass suicide in California in 1997, but there has been no damaging earthquake, as of this writing.)

In late 1811 and early 1812, however, there were floods on the Ohio and

Mississippi Rivers, followed by drought, "unprecedented sickness" among the inhabitants of the river valleys, strange weather and animal behavior, earthquake lights and sounds, an unusually brutal murder, Indian troubles, war looming with the British, and a fire-belching dragon—the first steamboat on the inland waters was descending the two rivers.

In Congress it was called "a time of extraordinariness." Noting the succession of earthquakes, a Connecticut paper commented:

> The period is portentous and alarming. We have within a few years seen the most wonderful eclipses, the year past has produced a magnificent comet, the earthquakes within the past two months have been almost without number and in addition to the whole, we constantly "hear of wars and summons of wars." May not the same inquiry be made of us that was made by the hypocrites of old: "Can ye not discern the signs of the times."

Many years later the writer Eudora Welty drew Robert Penn Warren's attention to accounts of that time. Warren noted in his epic poem *Brother to Dragons*:

> *And now is a new year: 1811*
> *This is the* annus mirabilis. *Signs will be seen.*
> *The Gates of the earth shall shake, the locked gate*
> *Of the heart be struck in might by the spear-butt.*
> *Men shall speak in sleep, and the darkling utterance*
> *Shall wither the bride's love, and her passion become*
> *But itch like a disease: scab of desire.*
> *Hoarfrost lies thick in bright sun, past season,*
> *And twitches like wolf fur.*

The poet graphically described the ax murder of a young African American slave by two of Jefferson's nephews, brothers named Isham and Lilburne Lewis.

As if ordained by a magus and in sync with the beat of the earthquakes, the murder was committed seventy-five miles east of New Madrid on the night of December 15–16 and discovered when the chimney in which the charred bones were hidden was toppled by the greater temblor of February 7. It was a remarkable confluence of violent events. Warren asked why:

> *Yes, God shook out the country like a rug,*
> *And sloshed the Mississippi for a kind of warning—*
> *Well, if God did, why should he happen to pick out*
> *Just Lilburne's meanness as an excuse?*

The most spectacular effects of the earthquake were experienced on water.

The noisy, belching steamboat, built in Pittsburgh and painted sky blue with a single paddle wheel attached to its stern, was more than one hundred feet long. On board the *New Orleans* was Nicholas J. Roosevelt, an associate of Robert Fulton and Robert R. Livingston. They had pioneered steamboat navigation on the Hudson River four years previously. Roosevelt was accompanied by his family, a pilot, an engineer, a crew of six, assorted domestic servants, and a large dog. His wife, Lydia, was pregnant and gave birth shortly before the earthquakes.

It was a demonstration voyage. Roosevelt chugged down the Ohio River and then backtracked to prove to skeptics that the vessel could navigate upriver as well. To emphasize speed, he proceeded at night—visibility permitting.

The clanking and clattering of the machinery, the various exhalations of steam, and the lurchings of the ungainly vessel obscured what was happening around it. Stopping to load coal in mid-December, they were asked if they had heard strange noises, seen lights, or observed the earth shake. No, they hadn't.

Roosevelt's account of the voyage was related by Charles J. Latrobe in *The Rambler in North America*:

> Hitherto nothing extraordinary had been perceived. The following day they pursued their monotonous voyage in those vast solitudes. The weather was observed to be oppressively hot; the air misty, still, and dull; and though the sun was visible, like a glowing ball of copper, his rays hardly shed more than a mournful twilight on the surface of the water.

Evening drew nigh, and with it some indications of what was passing around them became evident. And as they sat on deck, they ever and anon heard a rushing sound and violent splash, and saw large portions of the shore tearing away from the land and falling into the river. It was, as my informant said, "an awful day; so still, that you could have heard a pin drop on the deck." They spoke little, for every one on board appeared thunderstruck. The comet had disappeared about this time, which circumstance was noticed with awe by the crew.

The trees swayed wildly on the riverbanks, although there was no wind; others lay freshly uprooted in the water. The precipitous banks slid into the river, in one case inundating a flatboat and raft from which the occupants had fled. Islands known to the pilot were no more, channels were greatly altered, and the *New Orleans* became lost in the dark. They tied up to an island and stayed awake all night "listening to the waters which roared and gurgled horribly around them."

At dawn they discovered that they were near the junction of the Ohio and Mississippi Rivers, and by noon the vessel stopped at New Madrid. "Here they found the inhabitants in the greatest distress and consternation; part of the population had fled in terror to the higher grounds, others prayed to be taken on board, as the earth was opening in fissures on every side, and their houses hourly falling around them," noted Latrobe.

They took no passengers on board—the inhabitants, on second thought, distrusted the belching ark more than the trembling earth—and proceeded down the "fearful stream" to Natchez, where the steamboat arrived in early January. They had reached the end of their historic voyage "to the great astonishment of all, the escape of the boat having been considered an impossibility."

There was no accurate death toll, the region being too remote and communications too unreliable; but it is certain that more victims perished in the water than on land.

The mighty river was a mass of logs, disappearing islands, bubbling and churning waters, and for a short time it seemed to run backward. Latrobe, a self-described English gentleman who enjoyed wandering about the world, rather colorfully described the reversal of the river's flow. He wrote:

And finally, you may still meet and converse with those, who were on the mighty river of the West when the whole stream ran towards its sources for an entire hour, and then resuming its ordinary course, hurried them helpless on its whirling surface with accelerated motion towards the Gulf.

Vast tracts of land either sank or were uplifted by the powerful earthquakes. The Tiptonville Dome bulged upward some fifteen or twenty feet in the early-morning hours of February 7. The dome extended fifteen miles from just above New Madrid south along the Mississippi to Caruthersville. It was sufficiently high "to raise the surface above the reach of the highest floods," according to a USGS study published one hundred years after the earthquake.

The dome formed a temporary plug. The river backed up and flooded the surrounding countryside, giving rise to the popular belief that it had reversed its course. The water then rushed back into the riverbed and began to cut through the soft mud and sand of the uplifted dome, forming two sets of dangerous rapids. The booming noise of the rapids could be heard in nearby New Madrid.

Boatmen came upon this fearful spectacle unaware.

Mathias M. Speed noted a slackening in the normal river current due to the ponding of the river some distance above the plug. The current then increased greatly, Speed said, and the two boats that were lashed together were caught "in the suck." The men were frightened by "the appearance of a dreadful rapid of falls in the river just below us." The two boats made it through; others were not as lucky. The rapids lasted several days, by which time the river had cut through the soft dome and reached its normal level.

A Scottish botanist, John Bradbury, was descending the river on an oar-powered boat loaded with lead. Thomas Jefferson had given Bradbury a letter of introduction, referring to him as "a man of entire worth & first order." The crash of trees on the banks, Bradbury noted, "mixed with the terrible sound attending the shocks, and the screaming of geese and other wild fowl, produced an idea that all nature was in a state of dissolution."

· · · ·

The behavior of humans and animals differed from the norm on land.
A New Madrid resident noted after the December quake:

> It is really diverting, or would be so, to a disinterested observer, to see
> the rueful faces of the different persons that present themselves at my
> tent—some so agitated that they cannot speak—others cannot hold
> their tongues—some cannot sit still, but must be in constant motion,
> while others cannot walk. Several men, I am informed, on the night of
> the first shock deserted their families, and have not been heard of since.

The naturalist John James Audubon was returning on horseback to his
home in western Kentucky, some 150 miles distant from New Madrid. He
was traveling with a New Orleans merchant, Vincent Nolte. They went their
separate ways at Lexington, and Audubon rode across the Kentucky barrens.

"A sudden and strange darkness" rose from the western horizon, he
wrote in his journal. (Audubon wrote November, but he probably meant
January 23.) There was a calm, like the stillness before an approaching summer thunderstorm. He heard a sound like "the distant rumbling of a violent
tornado."

Audubon spurred his horse, but "the animal knew better than I what was
forthcoming and instead of going faster, so nearly stopped that I remarked
he placed one foot after another on the ground, with as much precaution as
if walking on a smooth sheet of ice." The horse hung its head, spread its four
legs wide, and groaned continually.

The bushes and trees shook, and the ground rose and fell "like the ruffled
waters of a lake." The "fearful convulsion" ended in a few minutes, the sky
brightened, and horse and rider galloped off. A daily succession of aftershocks continued for weeks. Audubon wrote in his journal of this time:
"Fear knows no restraint."

Audubon's traveling companion continued his journey. On January 23
Nolte's horse also balked "as if struck by lightning." The merchant, who was
an experienced traveler, boarded a boat. In company with twenty other
boats, it tied up at New Madrid on February 7. That night the remainder of
the raw frontier community was destroyed. Everyone, except a seventeen-
year-old girl with a broken leg who was left behind by her family, fled the

town site. The other boats cut their lines and were never heard from again. Nolte insisted on a daylight departure and made it safely to Natchez.

Others also reported unusual and strange phenomena. A New Madrid resident said of the early morning of December 16: "The heavens were very clear and serene, not a breath of air stirring, but in five minutes it became very dark." He was awakened by "a most tremendous noise." Five hours later, as the aftershocks continued, "the darkness returned, and the noise was remarkably loud."

Mathias Speed wrote of the sounds that he heard near New Madrid in the early-morning hours of February 7: "The constant discharge of heavy cannon might give some idea of the noise for loudness, but this was infinitely more terrible, on account of its appearing to be subterraneous." A man who had been marooned on an island in the middle of the river told Speed "that frequent lights appeared."

Besides the Tiptonville Dome, two other large mounds were thrust upward and numerous sinks were created. In fact, large sections of the central Mississippi River valley were known as "sunk country," particularly forty-mile-long Lake Saint Francis in Missouri and Reelfoot Lake in western Tennessee, where the land dropped between fifteen and twenty feet. "We supposed the whole country sinking, and knew not what to do for the best," said a frightened resident.

Over the years the stark outlines of dead tree trunks were reminders of the tenuous nature of the land, and on occasion fires raged through these dry ghost forests. Perhaps 150,000 acres of woods were destroyed by the earthquakes.

From huge sand boils came the innards of the earth. In one case, the ejecta consisted of the cranium of an extinct musk ox. It was donated to the Lyceum of Natural History of New York. There were sulfurous odors, perhaps dead organic matter or petroleum deposits. Sulfur odors were also associated with other earthquakes.

In what is regarded by modern seismologists as the best-documented account of the New Madrid earthquakes, Myron L. Fuller of the USGS confirmed the existence of most of the strange phenomena in his comprehensive 1912 report. Fuller combed the region by canoe and horseback for years and perused all the available earthquake literature.

Various causes were assigned to the earthquakes. The Territorial Assembly of Missouri blamed God, a botanist thought that a giant bed of coal or lignite had decomposed and collapsed, and a newspaper editor believed that a volcano was involved. Electricity, endowed with magical properties at the time, was also a candidate. Electrical fluids, it was thought, could cause near-simultaneous disturbances in different parts of the world. The electrical theory was given a boost in March 1812, when Caracas, Venezuela, was destroyed with a loss of more than ten thousand lives. In that year of violent earthquakes, there seemed to be a connection.

The aftershocks from the three huge quakes trailed off in the summer of 1812 and began to fade from memory. The upheavals of that time were not widely recalled until the late 1980s, when a prediction of another great earthquake again caused fear to flood the land.

FORT TEJON 4
(1857)

PALLETT CREEK

I set out from my home on an early February Sunday for the north slope of the San Gabriel Range, 380 miles to the south along the curving arc of the San Andreas Fault. The mountains had been thrust six thousand to ten thousand feet into the air by numerous earthquakes over millions of years and were dusted lightly with snow on this blustery day.

On the other side of the mountains lay the high-tide mark of southern California's population. Some twenty million people south of the fault tried not to think about earthquakes, although it had only been two years since the last one.

I drove south on the Antelope Valley Freeway and turned off at Sierra Highway, just beyond Palmdale. From there I wended my way eastward to Cheseboro Road and parked beside the California Aqueduct.

The morning was cold, but there were a few fishermen from the city sitting on stools beside the gleaming concrete. The water had been transported from the northern tip of the Sierra Nevada and was destined for southern California. The sinuous aqueduct took the line of least resistance and snaked along the fault line.

All the life-support systems—such as the aqueduct—that plug southern California into the rest of the state and the nation either cross or run parallel to the San Andreas Fault zone along a broad arc from north to west of the populated areas of Los Angeles and Orange Counties. These basic necessities include Feather River water, Colorado River water, water from the Owens Valley, freeways, lesser roads, railways, tunnels, natural gas and petroleum pipelines, electrical transmission wires, and communication lines.

One massive upheaval in the range of a magnitude 8 earthquake along the zone of fracture—such as was experienced in 1857—would immediately sever southern California from the outside world and leave only coastal roads, assuming they were passable, as escape routes.

The great vulnerability of the region was brought home to me when I later stood on the fault line above Interstate 5 at Tejon Pass, State Route 14 at Palmdale, Interstate 15 at Cajon Pass, and Interstate 10 at San Gorgonio Pass. I felt the tremors in the earth as the heavy traffic—over one hundred million vehicles a year on those four freeways—passed by on the stained concrete.

These ribbons of steel and concrete that have been imposed upon the fractured landscape look quite substantial, but actually they are fragile constructs that would crumble during a seismic cataclysm. The population would then be imprisoned by the faults that had thrust the surrounding mountains into the air, and the apocalyptic scenarios for the region on the verge of the millennium would be fulfilled.

We learn to ignore or not dwell for long upon such thoughts in California. History tells us that our chances are pretty good for slipping by such a major disaster.

I drove east from where I had parked by the California Aqueduct and found the Pallett Creek site without any difficulty, having been told its location by Kerry Sieh (pronounced See) when I visited the lean, athletic-looking Caltech geology professor in his office. Sieh was a graduate student at Stanford University in the mid-1970s when he began his search for past earthquakes. With help from his wife and his brother, Sieh worked the banks of Pallett Creek as carefully as a gold miner sifted his diggings.

The result for Sieh was gold of a different sort: a thesis for a doctor's degree and numerous articles in technical journals that brought him fame within the small world of seismology and its offshoot, paleoseismology. Since then, Sieh and others have dug numerous trenches across the San

Andreas and other faults in what has become a cottage industry. The original site is now a station of the cross for geology classes.

The South Coast Geological Society's guidebook that I carried stated, "The work by Kerry Sieh at Pallett Creek has proven to be a landmark study that has established for the first time the history for multiple earthquakes along a fault based on geologic evidence, and gave Californians a glimpse of the true risk from the 'Big One.' "

I walked down the path. The recent rain had erased previous footprints, and there were no other visitors that morning. It was a peaceful place and, like much of the terrain along the fault line, beguilingly beautiful.

The willow and cottonwood trees were bare, but a green fringe of watercress framed the tinkling stream, fed at this time of year by the snow on the nearby mountains that formed a backdrop to the south. To the north was the beige-colored landscape of the Mojave Desert and the prickly outline of Joshua trees. The freshness of the scene would be replaced in a couple of months by a dusty, crackling dryness that would last through the hot summer and early fall months.

I stood on a gently sloping earthen ramp that led to the creek. High flows in the stream after the turn of the century had cut a thirty-five-foot-deep gorge. What was partially exposed by nature and then deepened by earth-digging machines and shovels were various layers of stream-bed deposits dating back to 500 B.C. The thin dark lines of peat came from freshwater marsh plants. The marshes were covered periodically by floodwaters. Silt, sand, and gravel were laid down on top of the peat.

Gradually a record was constructed, a record that Sieh read by radiocarbon-dating the various peat formations that contained the swirls of seismic fingerprints. He published his first results in 1978 and 1984; and then, using improved dating methods, published a third set of figures in 1989.

The earthquakes of 1812 and 1857 that showed up in the trench were known from historical records. The dates for unrecorded earthquakes were 1480, 1346, 1100, 1048, 997, 797, 743, 671, and two other temblors that occurred between 529 and 46 B.C. The median dates were not as precise as the figures indicated because there was a sizable margin for error in the dating method.

The average interval between the ten most recent events was 132 years; the implication was that there would be a large quake on the southern portion of the San Andreas Fault around 1990. A more important finding was that the ten quakes were grouped in clusters. Four clusters contained two or three events with several decades between them in each cluster and two or three centuries between clusters. Recently the clustering of earthquakes has also been noted in Iran, Italy, Greece, Turkey, New Zealand, China, and Japan.

Sieh wrote:

> If this pattern continues into the future, the current period of dormancy will probably be greater than two centuries. This would mean that the section of the fault represented by the Pallett Creek site is currently in the middle of one of its longer periods of repose between clusters.

If there is one thing earthquake history teaches, however, it is lack of consistency and great uncertainty. Earthquakes have a nasty habit of not occurring at or near average, regular, or predicted intervals.

The record from which historical data has been extracted is less than infinitesimal. In terms of the thirty-million-year history of the San Andreas Fault, a paleoseismic record for a scattering of places exists only for a few thousand years. Written accounts of earthquakes date back to 1769 in California. A fairly complete seismographic record exists for the last seventy years. This record is much too thin a sampling on which to base firm conclusions, be they precise numbers or percentages. What is certain is that there will be more earthquakes, and a few of them will be large.

There were other unknowns: Was Sieh's work complete, meaning did he miss a quake or two, and was Pallett Creek a representative site? At most it would be representative of one section of the San Andreas Fault. The more southerly, central, and northern portions seemed to have different histories. Then, there were other surface faults with different names and records of fracturing, along with invisible faults—both known and unknown. The complexities were boundless.

"A HORRIFYING EARTHQUAKE"

The written earthquake history of California began in 1769, when the first Spanish explorers marched across the Los Angeles plain as it shook like jelly for one week.

The first overland expedition departed from San Diego on its way toward Monterey Bay on May 15. It was led by Gaspar de Portolá, a military man who would become the first Spanish governor of California. Also in the party were Miguel Costansó, an engineer and cartographer, and two priests, Juan Crespi and Francisco Palóu. All left journals that described the earthquakes. Portolá's account was briefest and to the immediate point, as befitted a military man. The man of science and the priests were wordier.

The pack train lurched northwestward, paralleling the coastline and the San Andreas Fault. Ahead went the scouts, followed by Portolá, his officers, and the two priests. The summer days were hot, so the marches were short. Fires, floods, and earthquakes, the latter achieving mantralike proportions, were mentioned in their journals. At the same time the Spanish repeatedly remarked on the extreme beauty and fecundity of the place.

Friday, July 28, 1769: The explorers halted on the southeast bank of the Santa Ana River opposite a large Indian encampment in what would become known as Orange County. They exchanged gifts with the Indians—beads and a silk handkerchief for seeds and a shell necklace. The bed of the river was thick with alders, willows, sycamores, and other trees they did not recognize. The land looked rich and easily irrigated. They could see evidence of flooding from the previous winter.

Portolá wrote:

> Here, at twelve o'clock, we experienced an earthquake of such violence [indecipherable in the original manuscript] supplicating Mary Most Holy. It lasted about half as long as an *Ave Maria* and, about ten minutes later, it was repeated though not so violently.

Proving again that "facts" are illusory and what occurs is in the eye of the particular beholder, the priests had a somewhat different take on the earthquakes. One wrote:

> I called this place the sweet name of *Jesús de los Temblores*, because we experienced here a horrifying earthquake, which was repeated four times during the day. The first, which was the most violent, happened at one in the afternoon, and the last one about four. One of the heathen who were in the camp, who doubtless exercised among them the office of priest, alarmed at the occurrence no less than we, began with frightful

cries and great demonstrations of fear to entreat heaven, turning to all the winds.

"Terrible" and "fearful" were alternative translations of the adjective used to describe the shock.

There is little doubt that the first shock was an earthquake and what followed for the next six days were aftershocks, since the strongest language—which in this case serves as a seismograph—was reserved for the original event. Sieh and the USGS estimated that the temblor was equivalent in magnitude to the most powerful southern California earthquakes experienced in the nineteenth century. If so, then deaths and damage in modern-day Los Angeles and Orange Counties would have been enormous from such an earthquake.

Sunday, July 30: Costansó noted a small earthquake in the morning, but the priests did not mention it. They camped later that day on the San Gabriel River near the site that would become the mission, and, the priests said, "In the afternoon we felt another earthquake." They gave the name of San Miguel Arcángel to the valley. When Junipero Serra, the founder of the chain of California missions, arrived at the site he renamed it El Valle de Los Temblores.

Monday, July 31: They fought their way through thick brush and low trees, failed to make good time, and camped near El Monte. "At half-past eight in the morning we felt another earthquake," the priests said. Costansó added the adjective "violent."

Tuesday, August 1: They rested on this day. The priests said mass and the soldiers received communion. "At ten in the morning the earth trembled," the priests remarked. "The shock was repeated with violence at one in the afternoon, and one hour afterwards we experienced another." Portolá counted "six or seven severe earthquakes" that day.

Wednesday, August 2: They camped near a river. "On this day we felt four or five earthquakes," said Portolá. The priests described the place as "a very spacious valley, well grown with cottonwoods and alders, among which ran a beautiful river." It was the Los Angeles River, now encased in concrete. They reported, "Here we felt three consecutive earthquakes in the afternoon and night."

Thursday, August 3: They passed through downtown Los Angeles on their way to the Wilshire District. "This afternoon we felt new earthquakes, the continuation of which astonishes us. We judge that in the mountains that run to the west in front of us there are some volcanoes." They passed the La Brea tar pits, which were spewing a lavalike substance. Portolá called them "extensive swamps of bitumen." He added, "We debated whether this substance, which flows melted from underneath the earth, could occasion so many earthquakes."

Portolá and his entourage, unknowingly following the line of the Hollywood Fault, continued on their way past Beverly Hills and then turned toward the north and the San Andreas Fault. There were no further reports of significant earthquakes in southern California until 1812.

FROM SAINT ANDREW'S ABBEY TO WRIGHTWOOD

Continuing in an easterly direction, I drove a short distance down Pallet Creek Road and turned into the driveway to Saint Andrew's Abbey. The signs stated "No hunting, except for peace" and requested that headlights be turned off at night. That was the time for silence and contemplative prayer on the edge of the rift zone.

The Benedictine monks were aware of the fault. The older ones had watched Sieh's progress and talked with Levi Noble, a neighbor and geologist. Although Noble was one of the first proponents of substantial lateral movement on the San Andreas Fault, the monks' clearest memory of him was his cigar-smoking wife.

The religious community was founded in China by the Abbey of Saint Andre in Bruges, Belgium. Saint Andre, Saint Andrew, and San Andreas are the same apostle. The fact that the fault and the monastery have the same name is a coincidence, say the monks.

When the Communists took over Chengdu, the capital of Sichuan Province, the monks and nuns were banished from the country. They arrived in California in 1952 and set about looking for a new home, preferably in a stable setting.

The monks purchased the fruit ranch in 1955 because of its tranquillity. Green lawns, a large pond, and tall trees in a desert setting with towering mountains as a background were reminiscent of an oasis in the Holy Land. Hidden Springs Ranch on the edge of the fault line was converted into a religious sanctuary and a profitable business enterprise. The stable became the modern chapel, the cow barn was made into living quarters for the monks, and the blacksmith shop is the home of the ceramics business.

When I checked into the retreat office, Brother Jonathan said, in reference to my query about earthquakes: "We are all full of seismic energy here, and we certainly have our faults." He added that they not only prayed to Saint Andrew, but also to Saint Emigdio, the Roman saint evoked for protection against earthquakes. When I told him a creek, peak, mountain range, canyon, and mesa seventy miles to the west were named San Emigdio, he was surprised.

The presence of the fault, which crosses the southern end of the 760-acre monastery grounds, meant little in terms of the daily life of the monks. They were, however, disturbed by the low-flying jets.

One little-known benefit of the fault, as determined by the military, is that it serves national security needs. Residents from Wrightwood in northeast Los Angeles County to Pine Mountain in Kern County mentioned the low-level flights. The military confirmed them.

Jet aircraft from various military bases fly as low as two hundred feet above the fault line at speeds in excess of five hundred miles per hour. Some are practicing contour flying. Others are playing a cat-and-mouse game, seeking to avoid radar detection within the shallow depression so that they can score a surprise attack on nearby Edwards Air Force Base. I heard that a six-pack of beer went to the winner of such contests.

The jets fly through the rift valley in pairs from west to east during daytime and nighttime hours, with afterburners screaming to give them the needed thrust over the low passes. They are so low that residents living or working on hillsides can see the color of the pilots' helmets. The shapes of some planes are quite radical, indicating the testing of such prototypes as the stealth bomber.

Scheduling for the use of the low-level training runs is handled at Lemoore Naval Air Station in the nearby San Joaquin Valley. The schedulers and pilots don't know it as the San Andreas Fault run. To them it is a route designated VR-1257 on their charts.

People, pets, livestock, and praying monks are frightened or disturbed by the noise. There have been complaints. The English author and mystic Aldous Huxley once lived in the area. He wrote:

> The stillness is so massive that it can absorb even jet planes. The screaming crash mounts to its intolerable climax and fades again, mounts as another of the monsters rips through the air, and once more diminishes and is gone. But even at the height of the outrage the mind can still remain aware of that which surrounds it, that which preceded and will outlast it.

All was quiet and serene during my brief stay at the monastery. That the meditative monks who "try to encounter, in full consciousness, the mysterious power hidden beyond the world of man and nature" were located in the fault zone seemed quite appropriate. I wished them luck and continued my journey along County Highway N4.

Valyermo is a post office, a Forest Service ranger station, and a ranch. The ranch, which shipped fruit under some famous brands in its day, was for sale. In 1933, the ranch owners and hired hands were eating dinner when the Long Beach earthquake struck far to the south. They ran outside. Structures shook, dishes rattled, and a few broke. It seemed for a few moments that the stone ranch house would collapse, but it held.

That afternoon I camped in the nearby Forest Service campground at Sycamore Flat, close by Big Rock Creek. I walked to the creek and lay down in the fresh grass. When I opened my eyes after a short nap, they happened to alight on a message carved above me in the bark of a tree: "Fuck you." That, too, can be a reaction to seismicity.

Big Rock Creek, formerly named Rio del Llano, was a potential source of water for the nearby Llano socialist colony. The Llano del Rio Cooperative Colony constructed a small dam in 1916 that trapped another supply of water in a sag pond on the fault line. In this manner Jackson Lake was formed.*

Local ranchers wanted no part of "socialist plunderers." They prevailed, and the colony was denied water rights from Big Rock Creek. That was one

---

* A sag pond is a depression formed by faulting in which water collects.

reason why the colony failed. The socialists departed in 1918, but Jackson Lake remains. It is an idyllic place for the masses to picnic and fish.

The next morning there was a couple in the neighboring campsite who had arrived late the previous night. As I was folding my tent, the man invited me over for coffee. I declined, suspicion being the initial reaction in the urban atmosphere that has seeped, like smog, across the mountains from Los Angeles.

I then thought better of my unfriendly response and took them a peace offering in the form of a paperback book that I had finished. They were a bit puzzled by what to do with the book, but the gesture broke the ice. We talked. The tenor of the conversation was that there was no escape.

They had been comfortably settled at another campground farther up the canyon when some city youths arrived with guns, liquor, and loud music. The couple quickly threw their camping gear into the back of their pickup and departed.

In a way, they were typical of their time and place. He was from inland Fontana; she was from coastal San Pedro. It was a freeway romance. He was of Italian and she was of Asian descent. Crime and earthquakes were getting to be too much for them. He was thinking of returning to his native upstate New York.

I explained that we were camped between the San Andreas and Punchbowl Faults. He said it really didn't make much difference because there are so many faults in southern California. The woman didn't say much as she stirred the campfire and sent dense clouds of smoke into the air.

They were arguing about that as I departed.

East of Big Rock the county road and fault line gain altitude; both reach their apex at the 6,862-foot level of Big Pines. Just before the series of sharp curves that climax at Big Pines is Appletree Campground. A dirt road heads east from the campground to one of the most dramatic expressions of the San Andreas Fault.

The water from Mescal Creek has added its erosive power to the gnashing of the earth and cut a deep, V-shaped canyon whose steep sides were composed of finely ground gray rock. There was a badlands aspect to the landscape. Trees tipped downward, canted at an angle by the slip of the land. Fog poured over the backside of the San Gabriel Mountains, periodically obscuring the upper reaches of the canyon.

I felt isolated and vaguely endangered.

I hesitated at the water tank, then continued to walk along the road, which was bordered on the north by a low ridge. The dirt road came to an abrupt halt. A side canyon that intersected the main canyon at a ninety-degree angle had wiped out the track. I followed a faint trail down into the side canyon and then bushwhacked up the main creek bed to the twisted remains of a steel check dam. My guess was that it had been erected to halt the erosion that was threatening the roadway far above me.

The sides of the canyon teetered on the edge of the angle of repose. Darkness hastened by the damp fog was descending. The disturbed landscape seemed to be closing in around me like a vise. I did not want to be pinned in that place and departed quickly.

When I reached my vehicle I dug out the South Coast Geological Society guidebook and read: "Just before the end of the road, on the left (north) side, is a ridge of crushed granitic rock that was pulverized and uplifted by shear and compression along the San Andreas fault. This was probably formed during the 1857 quake."

At the highest point on the fault line, one of two stone towers that had once been connected by a pedestrian overpass remained at Big Pines to mark the rift zone and the attempt by Los Angeles County supervisors to encourage outdoor recreation in the late 1920s and early 1930s. The remains of the tower and other cobblestone structures lie in the saddle between Blue Ridge to the south and Table Mountain on the north. The saddle is the fault line.

At the crest of the mountain is the Table Mountain Observatory, established in 1926 by the Smithsonian Institution and now run by the Jet Propulsion Laboratory (JPL) as part of the NASA network. From the 7,500-foot level of the mountaintop various optical devices look into distant space while the ground occasionally rocks and rolls beneath them.

Table Mountain is a tuning fork on the edge of the San Andreas Fault. Distant earthquakes are felt by observatory personnel and registered on the many seismic instruments planted there by various institutions. The mounting on one observatory instrument was fractured in one quake and minor cracks have appeared in the concrete. The buildings are specially reinforced.

On October 1, 1987, Dan Sidwell, the site manager, and one of his employees were standing on the asphalt on the southwest side of the sturdy garage-machine shop when the moderate Whittier Narrows earthquake struck, killing eight and causing extensive damage thirty-five miles to the southwest in Los Angeles.

Sidwell is among the few who have seen the earth move but leave nothing disturbed in its wake. A forty-year veteran of space work at JPL in Pasadena and on Table Mountain, Sidwell described his experience:

> We were standing there talking and suddenly we heard the fluorescent light fixtures and the metal in the racks that held all the extra stock tinkle and rattle. I said, "What the hell was that?"
>
> At that point I noticed the trees to the southwest of us shake, but none of the other trees were shaking. The next thing I knew was the trees close to me started shaking. Then I realized there was this wave in the ground traveling toward us and I watched it pick up the fourteen-by-fourteen-foot utility building that held all of the power and communications coming into the site. It shook that building that was twenty feet away, and then we watched it go past the telephone pole about ten feet away, shake it, and then it picks us up and sets us down.
>
> We turned around and watched it travel across the floor of the garage and shake all the things in there. There is a gas pump outside the machine shop. We watched it pick it up and set it down. There are two ninety-foot telephone poles out there. We saw them standing still and all of a sudden the tops of them leaned to the northeast and then they came back to the southwest and just did an upside-down pendulum routine. We watched it travel past the radio astronomy building, and we saw the tops of other trees shaking on down the ridge.

The wave disappeared in the direction of the Mojave Desert. Sidwell said it resembled a small water wave that never crested. He estimated it was from four to six inches high and had a base of eighteen to twenty-four inches. The surge traveled slowly, perhaps taking five seconds to cross the twenty feet from the utility building to the two standing men—the speed of a casual bicyclist. It was accompanied by "a low-key roar, a rumble," said Sidwell.

Earthquakes were not a new phenomenon to Sidwell. He is a native Californian and had experienced a number of them. In 1971 he lived fifteen miles from the epicenter of the San Fernando Valley quake. Sidwell walked outside and saw the dust rise from the San Gabriel Mountains, which had just been given a good shake.

Of his experience on Table Mountain, Sidwell said, "I was in awe of what I saw coming toward us, moving by, and traveling down the mountain." There was no damage at the observatory. "It didn't leave a telltale mark anywhere," he said.

There are other accounts of such waves. A scientist on the staff of the Mount Wilson Observatory watched a cement floor rise four to six inches during a 1918 earthquake and then subside. He fled the Pasadena laboratory. A subsequent report by the National Research Council stated:

> Immediately after the cessation of the shock he returned and inspected the floor which showed no (new) cracks of even minute dimensions! Moreover, many relatively unstable objects on the tables and shelves in the laboratory remained in position, apparently undisturbed. It is plain, then, that in this critically observed case the apparent deformation of the floor resulted largely if not wholly from the observer's psychologic reaction to motion communicated to his body, or eyes, by the earth shaking.

A number of examples of such waves were cited in a recent discussion on the Internet news group sci.geo.earthquakes. People reported having seen such waves in Greece, in New Zealand, and on the West Coast. A woman described what she had witnessed while a junior high school student in Tacoma, Washington, during a 1965 earthquake. She said, "It is really refreshing to hear other people reporting this same sensation; I was beginning to wonder if it was only a psychological experience."

Seismologists, who would rather not deal with intangible phenomena, tend to downplay these reports. They credit slow-moving surface waves, resonance caused by passage of a seismic wave, or vivid imaginations. There is no clear explanation for such waves because they are hard to trap, and there is little interest in doing so.

Below in his home in the mountain community of Wrightwood, through which the fault passes, Sidwell has felt numerous minor quakes. "This place jiggles constantly," he said.

The principal fear of Wrightwood residents is a temblor in combination with an intense period of rainfall or a fast-melting snowpack. Such conditions could trigger a landslide from Wright Mountain. The mountain, with a visible gouge in its flank, hangs over the picturesque village like the sword of Damocles.

The sword dropped in 1941.

The blue-gray Pelona schist had been ground to a floury consistency over millions of years by tectonic motion. It became the consistency of wet cement when the snowpack melted rapidly late that spring. Residents thought there was an earthquake. In pulsing waves congealed dirt, pebbles, boulders, trees, highway signs, and crushed structures surged past at speeds up to fifteen feet per second. The ice-laced mass—the antithesis of hot lava—splattered onlookers. The debris flow petered out five thousand feet lower and fifteen miles out in the desert.

The rift zone passes on an east-west axis through older homes and vacant lots where building is no longer allowed on Lone Pine Canyon Road and continues on through the country club swimming pool, Our Lady of Perpetual Snows Catholic Church, and the remnants of a sag pond once called Wrights Lake. The fault zone travels through the base facilities of Mountain High Ski Area at the foot of Table Mountain.

In town I heard stories about lots that were purchased by buyers who then found they were unable to build upon them. I was assured by real estate salespeople that there was full disclosure of the presence of the fault. One sales tool was a large map reproduced from the *Los Angeles Times* that showed the many faults throughout southern California—the message being that it was more likely to happen elsewhere than here.

Wrightwood is at the western end of Swarthout Valley, whose extension toward the east is Lone Pine Canyon. It was at Lone Pine Canyon that *The WPA Guide to California* (1939) chose to deal with the fault, unaccountably stating that it ended in the Caribbean Sea. Near the top of Lone Pine Canyon three researchers, one of whom was Sieh, determined that a large pine tree lost its crown between the end of the 1812 growing season and the start of the 1813 season. That was one indication the 1812 quake was centered along this section of the fault.

The scientists also measured the tree-ring growth of nine pine trees in the Wrightwood area and concluded that they were disturbed during the same period of time. That the early-December earthquake struck here, yet its most devastating effects were felt far to the south at San Juan Capistrano, was testimony to the fragility of distant structures, the vicissitudes of soil and rock, and the capriciousness of seismic events—all of which counts for more, in terms of casualties and damage, than proximity to the fault line.

From the clarity of the mountainside village, the descent down Lone Pine Canyon was into a light gray veil that became the dense smog seeping over Cajon Pass from the Los Angeles Basin. The abrupt plunge seemed like the approach to a ghost city situated in hell.

At the bottom of the grade is Lost Lake, another sag pond. The lake is a popular place to fish and to dump bodies murdered elsewhere. No one was having much luck the day I was there, perhaps because some tough-looking gang members were throwing rocks into the small lake bordered by reeds. A thickly bearded man with a large dog was muttering about setting off some dynamite to stun the fish.

It was not until 1902 that a geologist recognized that the terrain had been severed by a fault that four years later came to be known as the San Andreas. Caltech seismologists recently excavated a trench near here and determined that at this point the San Andreas had ruptured six times in the last one thousand years.

The lake that formed in the depression created by faulting was fed by springs. On occasion helicopters scoop water from its surface to fight brush fires along nearby Interstate 15, the main Los Angeles–Las Vegas artery.

All the sounds hinted at movement. Ducks splashing in the lake and the crackling of electricity in nearby high-voltage transmission lines were superimposed upon the more distant moan of a train whistle. The hum of freeway traffic emanating from where the San Gabriel and San Bernardino Mountains are separated by the narrow pass floated along the fault line.

I drove down a remnant of fabled Route 66, which parallels the interstate. A bronze plaque set in a concrete obelisk on the fault line at Blue Cut noted that Cajon Pass was "an important natural gateway."

Besides the constant flow of traffic on the eight-lane freeway, two sets of railroad tracks convey four passenger and fifty freight trains daily. There are five high-voltage transmission lines, five fiber-optic telephone lines, three pipelines that carry petroleum products, two large natural gas pipelines, and a number of minor utility lines. Forest Service officials have voiced concern about the number of vital conduits jammed into such a vulnerable place, but nothing has been done to remedy the situation.

THE 1812 OVERTURE CONTINUES

The epicenter of the first of two December 1812 earthquakes was located along the Valyermo-Wrightwood portion of the fault. The powerful quakes—one on the eighth and the other on the twenty-first of the month—were the first warnings that multistory structures could not be designed and built in a haphazard manner in southern California. The widespread destruction of churches was a vivid testimonial to the architectural hubris that exists to this day in the state.

The year of 1812 was known as *el año de los temblores* in California. A series of tsunamis inundated the low-lying areas of San Francisco, and at the south end of the bay the church at the Santa Clara Mission was "cracked considerably."

The earthquakes in southern California began in May and were incessant for the remainder of the year. Hubert Howe Bancroft, the California historian, wrote: "A series of earthquake shocks, the most fatal if not the most severe that have ever occurred in California, caused this year the wildest terror throughout the southern part of the province."

John B. Trask was a medical doctor, an amateur geologist, and the first state geologist. He combed mission records and talked to inhabitants in the 1850s. Trask wrote of the earthquakes of 1812 in southern California:

> Their frequency was not less than one each day or two; four days seldom elapsing without a shock. As many as thirty shocks occurred in a single day on more than one occasion. So frequent were they, that the inhabitants abandoned their houses for the greater part of this period, and lived under trees, etc., and slept out of doors at Santa Barbara.

At Mission San Juan Capistrano, near the coast in what is now Orange County, there were forty days of earthquakes climaxed by a particularly hard tremor on October 21. Mortar fell from the vaulted nave of the mission church, and the 125-foot tower swayed precariously. The imposing edifice, designed in the form of a cross and with seven domes forming the roof, was built of unreinforced stone. The thick tower was the two-terrace type. There were four bells. Perched on top was a gilded cock.

The church rose toward the sky from a sere plain, was visible from ten miles away, and was the most magnificent and pretentious edifice in all of California—the equivalent in its time of, say, the Transamerica Tower in San Francisco.

Besides being a structural catastrophe just waiting to happen in known earthquake country, the church's construction was faulty. The master stone-mason died in 1803. The priests took over supervision of the Indian converts, known as neophytes, who did the actual work. The quality of the stonework on the final portion of the church, that being the nave, was inferior compared to what had been done before.

There were approximately fourteen hundred Native Americans at the mission, which was at the peak of its prosperity. They came from a variety of tribes in and around the area and could be viewed either as impressed slaves or happy converts. At the time, the priests at the mission saw the Indians as "poor and wretched; wherefore we find meekness and submissiveness to be their principal characteristics."

Indian men, women, and children hauled heavy stones to the site from a quarry six miles distant. When the church was completed in 1806, the padres proudly noted that it had been constructed by the Indians "at the cost of supplication and labors."

A small number of Indians were in the midst of their supplications on Tuesday, December 8, 1812. They were gathered for the early morning mass on the feast of the Immaculate Conception. A high mass, which all neophytes were expected to attend, was scheduled for later that morning. The Native Americans, most of whom were women, stood on the dirt floor of the nave while the *gente de razón*—the term Hispanics applied to non-Indians—were arrayed closer to the officiating priest, who conducted the service on the tiled floor.

The December morning was clear and "uncommonly warm," according to Trask. His account, published in various scientific journals a half century later, continued:

> About half an hour after the opening of service, an unusual loud, but distant rushing sound was heard in the atmosphere to the east and also

over the water, which resembled the sound of strong wind, but as it approached no perceptible breeze accompanied it. *The sea was smooth and the air was calm.* So distant and loud was this atmospheric sound that several left the building attracted by its noise.

The earthquake struck a few moments later. The priest fled out the side door of the sanctuary, followed quickly by the other Hispanics, none of whom died. The bodies of forty Indians were found huddled near a door in the nave that may have jammed. A church report stated that "only six escaped" but didn't say how. There was no record of the number of injured, if any.

Later reconstructions of what occurred have the tower falling away from the church and into the plaza and the nave toppling of its own accord, or the tower falling on the nave and the whole, heavy mass of masonry and stone descending upon the Indians. The back of the church, which was of better construction, remained standing, as did an adjacent one-story adobe structure known as the Serra Chapel.

These were the first recorded deaths from earthquakes in California. It is doubtful that there were many structural fatalities before this time. The Native Americans lived in much smaller, flexible, and forgiving shelters. But the Spanish churches toppled all over southern California during that month of sustained terror.

The churches at the San Gabriel, San Fernando, San Buenaventura (Ventura), Santa Barbara, La Purisma Concepcion (Lompoc), and Santa Ynez Missions were either damaged or destroyed. A Spanish vessel anchored off Ventura was damaged by a tsunami, a giant wave flooded the lower reaches of Santa Barbara, and a ship north of the settlement was thrown up and over the beach and then swept out to sea. There were reports of imagined volcanoes in the coastal hills.

Twenty years later the following conversation was recorded by a resident who dined with a priest:

> In speaking to Padre Luís Gil Taboada, he told me that in 1812 there had been very strong earthquakes at Santa Barbara while he was there. That on the [twenty-first] of December while he was at the presidio there occurred an earthquake so violent that the sea receded and rose

like a high mountain. He, with all the people of the presidio, went running to the Mission chanting supplications to the Virgin. I asked him,
humorously, why he had not gone to see if there were a ship at the foot
of the mountain of water. He also assured me that they had placed a
pole with a ball tied to it. It was fastened in the ground at a place where
the air would not move it, and that it was in continual movement for 8
days. After the 8 days the ball was still for 2 or 3 hours and then started
to move again, and this lasted for about 15 days.

This device may have been the first crude seismograph in California. The
epicenter of the December 21 quake was near Santa Barbara. Seismologists
would later assign high intensities to these two quakes.

The ruins of the church at San Juan Capistrano have been romanticized
over time by landscape painters, writers, and the swallow legend. They have
been compared to the ruins of Rome and Greece and called the "Alhambra
of America" and the "Melrose Abbey of the West."

The gloss masks the true significance of the ruins, which are visible to this
day.* John O'Sullivan, the priest who undertook the reconstruction of the
mission and became an authority on the stone church, wrote in 1912:

> The venerable crumbling walls have been studied and painted sym-
> pathetically by artists from near and far, measured with enthusiasm by
> architects, builders have stood in open-mouth admiration of the mas-
> sive concrete work done by the padres a hundred years before it dawned
> on the modern builder that the same, with steel reinforcement, was the
> proper mode for California.

FROM PALMDALE TO TEJON PASS

While the drive east from the Antelope Valley Freeway led to the epicenter of the 1812 quake, to the west was evidence of the great 1857 temblor.

I parked on Avenue S, just west of the freeway, and climbed the low hill to
the north. Below, the traffic moved unceasingly back and forth across the
fault line from the Mojave Desert to the populated coastal plain.

---

* They are a 186-year-old oddity, since the usual practice in California is to quickly erase any evidence of a natural disaster.

The road cut provides the most dramatic view into the interior of the fault. Here you stare into the very mouth of the beast and try to divine its inner workings. The afternoon light is a virtual spotlight on the west-facing side of the six-lane freeway.

Atop the bluff was the favored viewing spot for geology field trips. If any place on the entire fault line cried out for public access and a comprehensive interpretive exhibit, this was it. But natural disasters, or the potential for them, are not advertised in this state.

The most recent line of faulting is marked by an asphalt patch over the concrete pavement on the northbound on-ramp from Avenue S. The expansive nature of the clay that oozes from the fault warps the pavement. It is not the single strand of the San Andreas Fault, however, but rather the swirling turbulence of the entire fault zone splashed across the wide mural of the road cut that makes for such a dramatic representation of the chaotic earth.

The bared fault zone is a petroglyph of movement frozen in time. The lines of buff-colored sandstone and dark brown clay resemble an abstract cave painting. The flowing, twisting, parallel strata and the straight dark slashes join force to grace.

I looked below and spotted a yellow emergency phone powered by a solar panel. My imagination clicked into present time. What if I picked up that phone and reported that a huge earthquake was about to occur—the Big One, as it was known. Assume that I had special knowledge, and the 911 operator believed me. What could she do? Whom would she call? What could they send that would make any difference? A fire truck?

The fault gave as well as took away. Gypsum was present in the clay. The hill was one of California's earliest known sources of the mineral used in such plaster products as stucco. During the early years of the century, more gypsum was taken out of the hill than from any other single source in the state. Stucco replaced adobe as the favored exterior covering for homes in southern California, and fared no better in earthquakes.

Newly stuccoed homes surrounded me. Construction has been ceaseless in the booming Palmdale region in recent years.

The city of Palmdale, which encircles this site, is bold and brash and disheveled, as Los Angeles was before it reached a measure of equilibrium.

Like Los Angeles, this desert city suffered a bad press from outsiders. Three weeks after the Northridge earthquake of 1994 an article appeared in *The Wall Street Journal* under the headline "Prospects Look Bleak for California City: Palmdale Suffers from Foreclosures, Urban Woes." The first four paragraphs read:

> PALMDALE, Calif.—At 4:31 a.m. on Jan. 17, the fortunes of this already troubled city took a tumble.
>
> Twenty-five miles away, the Los Angeles area earthquake collapsed a five-story-high freeway intersection, severing Palmdale's main connection to the urban center where many of its now-rattled residents work.
>
> Suddenly, hours have been added to the commuting times of thousands, a stroke of misfortune for a city where things are already so bad that calls for Christmas charity baskets quadrupled this year at a local municipal aid agency.
>
> Though local economies are improving in much of the nation, the forces of Southern California's entrenched recession and damage from its earthquake are writ large here, with troubling financial and social repercussions.

The freeway has been repaired, traffic is rolling again, and the earthquake has been forgotten or, if remembered at all, is recalled as a time of inconvenience. True, a major computer manufacturer chose not to locate its only U.S. manufacturing plant in Palmdale because of its proximity to the fault, but there were compensations.

Palmdale boosters pointed out that, after all, the quake was centered in Los Angeles. Palmdale could therefore be considered a safer place. People bought that argument. Palmdale was the second fastest growing city in the country and the fastest growing in California during the first five years of the present decade.

Because the growth in Palmdale is so recent (from 12,000 in 1980 to over 100,000 in 1995) the fault zone, which cuts across the southern portion of the city in widths varying from less than five hundred to one thousand feet, is relatively uncluttered. State laws enacted the previous decade make it extremely difficult to build in an active fault zone. However, it is not only the San Andreas that could cause damage but also three of its branches in Palmdale: the Cemetery, Nadeau, and Littlerock Faults.

A state report on the fractured nature of the Palmdale segment of the fault line read:

> Without doubt, the San Andreas fault and many of the fault segments within the San Andreas Fault Zone are active faults capable of significant, and locally spectacular, surface disruption associated with faulting during future large earthquakes.

One possible result of a strong earthquake would be the rupture of the nearby California Aqueduct and the failure of the dam creating Palmdale Lake, which in combination or singly could cause widespread flooding.

I visited the lake, another dammed sag pond on the fault line, on a Saturday morning. The water comes from the state aqueduct. The lake is private, restricted to members of the Palmdale Fin & Feather Club. The local water district leases it to a concessionaire, that being one way to control access and gain revenue.

The signs stated the usual for such places: Private Property, No Trespassing, Members Only. I walked in and looked around. It was a larger version of the monastery grounds: quiet, peaceful, tranquil, shaded, green. There was a large expanse of sparkling water in a desert setting. An older couple was catching fish, a younger man wasn't. He wondered aloud what they were using for bait.

I asked the woman at the bait shop–store–reception center what were the advantages of membership. Members had caught 91,000 of the finest stocked trout last year, she said, and 40,647 had been caught as of mid-April this year, the season being from February to October. It occurred to me that if it hadn't been for the fault-formed lake, there would have been no trout fishing in the arid desert.

To the west of Palmdale on County Highway N2 there is a chain of smaller sag ponds that have been dammed and transformed into recreational facilities. Those seismically created bodies of water have attracted refugees to the 1857 section of the fault from earthquake and crime-ridden Los Angeles.

Mike Davis wrote of this place in *City of Quartz*:

> Setting aside an apocalyptic awakening of the neighboring San Andreas Fault, it is all too easy to envision Los Angeles reproducing

itself endlessly across the desert with the assistance of pilfered water, cheap immigrant labor, Asian capital and desperate homebuyers willing to trade lifetimes on the freeway in exchange for $500,000 "dream homes" in the middle of Death Valley.

Giant homes and monster mansions designed in the styles of faux Tudor, French château, southern plantation, or a combination of the aforementioned exotic architectural styles sat atop small hillocks that commanded views, much like fortresses above potential battlefields. They were walled off from the marauding masses by sturdy fences and No Trespassing, Private Property, Guard Dog, and Armed Response signs. Huge black gates rolled silently on electronic command across the gleaming black asphalt of new driveways that overlay the shifting surface of the earth.

A decrepit windmill stood above the ruins of a former working ranch served by a dirt road. A new play ranch was called "Pitchfork." The Leona Valley will absorb the next crunch of high-density development that is spreading out from Los Angeles. When I passed through, the land was being prepared for 7,200 new homes. The Ritter Ranch was billed as the largest residential development in progress at the time in Los Angeles County.

Throughout the Leona Valley there is ample evidence of massive faulting during the 1857 event. There are well-preserved benches, notches, troughs, and stream offsets measuring as many as twenty feet. More than one fault has ruptured here. "Numerous faults of various kinds are scattered across the alluviated terrain of Leona Valley, as well as along the slopes of the bordering mountains," according to a geology field guide.

From the Leona Valley and the immediate fault line I detoured a short distance south to the site of what is referred to as the greatest "man-made" disaster in the state's history. As with the term "natural" disaster, "man-made" lacks clarity. It is difficult to separate the natural and human components of such events.

The Los Angeles Aqueduct transports water from the Owens Valley on the east side of the Sierra Nevada to the San Fernando Valley, where it is distributed through part of Los Angeles. Three hundred feet under Elizabeth Lake on the San Andreas Fault, the aqueduct is a concrete tunnel that delivers water to two powerhouses in the nearby San Francisquito Canyon. For a brief period of time in the 1920s, the Saint Francis Dam and Reservoir was located between the powerhouses.

The dam was built under the direction of the legendary William Mulholland, who was responsible for bringing Owens Valley water south to assure the growth of Los Angeles. He envisioned a storage facility below the San Andreas Fault, should the fissure rupture or should angry Owens Valley ranchers, who felt they had been cheated out of their water, disrupt the supply. There were dynamite attacks on the aqueduct in the summer of 1927, and the dam was enlarged accordingly.

The base of the nearly two-hundred-foot-high concrete dam—the largest arched gravity dam in the world at the time—was knowingly built upon the visible San Francisquito Fault. Mulholland was asked at the coroner's inquest if he knew the dam was in a fault zone. He replied, "Well, you can scarcely find a square mile in this part of the country that is not faulty. It is very rumpled and twisted everywhere."

With the water lapping near the top of the dam's spillway for the first time, a leak at the junction of the fault and the concrete worsened on March 12, 1928. Mulholland, accompanied by his trusted lieutenant, Harvey Van Norman, inspected the dam and pronounced it safe in the morning.

It was a still night. A motorcyclist passed the dam on the country road and heard rocks falling. Nearby residents thought there was an earthquake. At 11:57 P.M. the lights flickered. The dam failed within seconds and twelve billion gallons of water—a gigantic "wall of death" more than one hundred feet high—poured down the canyon, destroying everything in its path.

Five hours and thirty minutes later the diminished flood reached the ocean, fifty-four miles distant in Ventura County. At least 420 people died and nine hundred structures were destroyed. It was the greatest civil engineering failure of the twentieth century in this country.

Mulholland was a ruined man. As with the padres at San Juan Capistrano, he was guilty of engineering hubris in a fractured landscape. A number of investigations were mounted, but none was very thorough. The same type of dam, only much larger, was being proposed for construction on the Colorado River. It was a time of massive water developments in the West; and the biggest, Hoover Dam, was before Congress for authorization. Los Angeles needed more water and power. That was what counted.

The exact cause of the failure of Saint Francis Dam—guesses ranged from sabotage to a slippage of the notorious Pelona schist, perhaps triggered by a discrete earthquake—has never been determined. Like the Bible, Charles F. Outland wrote in his book *Man-made Disaster*, "one can prove or disprove any point of view." An engineer, J. David Rogers, concluded: "Recent analyses reveal that the Saint Francis Dam was woefully inadequate in any number of potential failure modes."

Saint Francis Dam was the first in a series of catastrophes, near-failures, or damage to dams for seismically related reasons in southern California from 1928 to 1994. All but one of the dams was operated by the Los Angeles Department of Water and Power. Each successive failure led to stiffer standards for inspection and construction and made one wonder about the present adequacy of dam safety measures in earthquake country.

There is no monument to the disaster at the site of the Saint Francis Dam. One mile downstream at the restored Powerhouse No. 2 a commemorative plaque is affixed to a piece of the dam's concrete. It sits behind a chain-link fence and cannot be read from the road. The remnants of the dam were dynamited by the Los Angeles Department of Water and Power. The failure of a dam, like the presence of a notorious fault, was not something that was advertised in California.

The site was sprinkled with wildflowers. I climbed the embankment and made my way along the top, hopping from one concrete shard laced with rusted rebar to another. Huge chunks of concrete, some the size of a house, lay downstream, weathered to the point where they resembled boulders or the ruins of some ancient temple.

From the dam site I drove back up the canyon to the office for Powerhouse No. 1, where I had arranged to meet Dan Kott. I wanted to see the small museum in the powerhouse. We talked in the hallway and were joined by two other water and power employees, Jim Thomas and Carlos Gomez.

Five days after the January 17, 1994, Northridge earthquake Kott, Thomas, and two others were inside the underground aqueduct when a minor aftershock occurred near the epicenter of the quake, some twenty miles distant.

The four men were walking down the gradually sloping six-mile tunnel between the two powerhouses, checking for damage. Although usually a routine chore, this was the first time they had taken emergency supplies with them. They had no idea what they would encounter after the quake. They thought about the possibility of a cave-in as they slowly made their way through the eleven-foot-diameter concrete tube buried in the east wall of the canyon.

The men felt the dampness closing in around them as they sloshed through the puddles and mist in the humid shaft. Although most of the flow of water had been secured above them, there was still a trickle along the slippery floor. They wore hip boots. There was absolute darkness, except for darting headlamps and flashlights. They reached one of the few alcoves and paused.

Thomas said:

> It was a real shock as it came through. It sounded like a cannon shot. We could hear it go wooom, Wooom, WOOOM, building as it came toward us, and then passing us by and going WOOOM, Wooom, wooom. It was coming up canyon from south to north. You could hear it clearly, and you could hear it as it got to us, passed by us, and then traveled up the tunnel.
>
> So when we got out I called the engineers downtown and asked, "If that was an aftershock, how come there was no shaking?" And they said the deeper you go, the less movement there is; and as it ends up, you get more sound and maybe a little percussion as it goes by.

Kott added, "I had the awareness of something moving through the tunnel and actually passing us by. It was the only time I was ever afraid in there. But it was over before we knew it."

Gomez, the safety guard for the four men, had been sent on the morning of the quake to Fairmont Reservoir near the fault line to see if there was any damage. He said, "It was really bizarre. When I got there, the lake was like glass, except for the trout. They were just jumping right out of the water. Anytime there was a little aftershock, you could see the fish go up into the air."

The walk through the San Andreas Fault in the seven-mile-long Elizabeth Lake tunnel section to the north was always nerve-racking for Kott. "It is kind of an eerie feeling being within the earth at the point where those two huge tectonic forces meet," he said. There were cracks in the concrete-lined walls at the junction of the two plates and water leaked into the tunnel.

Kott was no stranger to earthquakes. As a teenager, he was living in a house that was shaken off its foundation during the February 7, 1971, San Fernando quake. His father worked for the department, and the family lived near the epicenter at Van Norman Reservoir.

The bridges over Interstate 5 collapsed a few hundred feet away. The Sylmar converter station, where 750,000 volts of electricity from the Pacific Northwest was changed from direct to alternating current, blew up in a tremendous white flash that Kott thought was the light from a thermo-nuclear explosion that had pierced the black curtains in his bedroom. And lower Van Norman Dam almost failed, forcing eighty thousand people to evacuate their homes in the San Fernando Valley.

After talking with the men in the hallway, I visited the museum, where the logbook of the powerhouse operator on duty when the dam failed was displayed in a glass case. It was opened to a handwritten entry made in red ink at 1:09 A.M., March 13, 1928:

> Spainhower Reports St. Francis Dam has Gone out. Lake is dry. Grade Caved in until it is impossible. Will go to Powerhouse 2 via Bee canyon tower line road.

Powerhouse No. 2 and the employees on duty had vanished under 110 feet of rapidly moving water, mud, and heavy debris. Miraculously, the heavy General Electric transformers that were yanked from the tracks they had been sitting on could be repaired. They are now bolted to the concrete foundation. The dings and weld marks in the thick steel sides of the cylindrical transformers are still visible.

West from the chain of sag ponds that have been transformed into magnets for recreational and residential developments was the tiny hamlet of Three Points, whose only public attraction was the small café. As I entered, a man cynically anticipated the question that was not on my lips: "Where are the poppies?"

No, I said, I was interested in the San Andreas Fault. Jim Bowman warmed to the subject. He had studied geology in college and had once tended seismic instruments for Caltech along the north slope of the San Gabriel Mountains.

As Marta ("call me the queen of Three Points") served coffee and break-fast, Bowman and others told me of the oak trees that grew only along the fault. Water was easier to find on the south side, they said. They also con-firmed the stories of low-flying jets. But they didn't know anything about the Neenach formation.

Just beyond Three Points on the Oakdale Canyon Road, the Neenach Volcanic Formation lies along the north side of the San Andreas. The matchup between these volcanic rocks and those on the opposite side of the fault 190 miles to the northwest at Pinnacles National Monument is the most visible proof of the tremendous lateral forces at work under California.

The two formations are the eroded and faulted remains of a single eight-thousand-foot-high volcano that erupted violently some twenty-three mil-lion to twenty-four million years ago. Then the Pinnacles formation began to move northwestward, leaving the Neenach rocks behind at an average rate of a half inch a year and considerably more in such years as 1857.

The Neenach formation is now hilly ranch land, comparable to the lower reaches of the national monument. Although the rock types, their stratigra-phy, and their chemical properties match, visually the spectacular upper por-tion of the Pinnacles formation seems more akin to the sandstone rocks at Devil's Punchbowl on the same side of the fault or, twenty-five miles to the east on the opposite side, the Cajon Formation (also known as the Mormon Rocks) close by Interstate 15.

Visual impressions, however, can be deceptive. Geologists at one time thought the Devil's Punchbowl and the Cajon sandstones came from the same source. They have since abandoned that theory because the fossil record doesn't match.

The remains of the ancient volcano and the absence of its other half, ripped from its insubstantial mooring and flung in very slow motion toward the northwest, drew me. I parked on Oakdale Canyon Road, which parallels the fault, and walked north on the Pacific Crest Trail, which led into the vol-canic formation and down onto the floor of the Mojave Desert.

There was a sturdy cattle gate, erected by the Tejon Ranch, over whose property the trail runs. Heat began to radiate upward as I left the oaks of the rift zone and entered the chaparral. Lizards darted across the seldom-used desert portion of the trail. I was alone in rattlesnake country, and no one knew I was there. Had I brought enough water?

At an overlook, I glanced back at the fault. It was a clearly demarcated line separating two different landscape provinces. From the cottonlike texture of the scrub vegetation rose the thick wool of the dense stands of pines above the road.

I retraced my steps, which were the only human marks in the volcanic soil, crossed the fault line, and climbed the well-trod path into the cool, thick greenness of the San Gabriel Mountains.

Below, I could see the dark red volcanic outcrops of the Neenach formation, which floated like an ancient reef upon a tossing sea. In the middle distance toward the west was Quail Lake and the California Aqueduct, which splits into two branches at that point. Farther away was Big Bend, where the east-west axis of the fault curves toward the northwest.

Fifteen miles in that direction the fissure crossed Interstate 5 at the crest of Tejon Pass, but the road cut was nowhere near as impressive as it was at Palmdale. Vegetation obscured the swirls.

I stood above the freeway and looked down upon the brake-check area for the huge trucks and trailers about to descend into the Los Angeles Basin. On flatbeds and in rectangular and round containers were oranges, tomatoes, kitchen appliances, toys, electronic exhibits, steel, new cars, wrecked cars, beef, baked goods, lumber, petroleum, and scrap metal. As the trucks paused at the end of the long climb, they shifted gears in the fault zone and began their downward plunge.

From Indian track, Spanish horse trail, stagecoach route, dirt road, two-lane then three-lane Ridge Route of the twenties and thirties, the four-lane Golden State Highway of the forties, and the impersonal interstate of the sixties and later, millions upon millions of people have crossed the fault unaware of its existence on this major north-south route.

I had no idea this was the fault zone when, a quarter century ago, I was the first to come upon the smoking remains of a tangled mass of cars and trucks at this same spot. Nothing human moved at that late hour as the smoke from the wrecks mingled with wisps of fog. There was little I could do, so I executed a slalom run through the wreckage and drove to the nearest phone to notify the highway patrol.

The prevalence of disasters is offset by stunning beauty in California.

The wildflower season was at its height, and poppies spread in wild

profusion over the softly molded ridges at the pass. The promotional literature I had picked up at a nearby coffee shop declared: "The golden poppy
has always symbolized California's good life. Its color mirrors California's
climate and charm, and the profusion of its blooms reflects the richness of
the Golden State."

A short distance to the north of the pass was the tastefully designed headquarters of the Tejon Ranch Company, the single largest private owner of
contiguous land in the state. A company attorney said that the ranch was
about to lease the property on which the Neenach formation was located to
a private hunting club. Quail, chukars, and imported pheasant would be the
principal game on the marginal grazing land.

Earthquakes?

He recalled that the old adobe headquarters of the ranch had been
destroyed in the 1952 Tehachapi quake, the most powerful in the state since
1906. The temblor was on the short White Wolf Fault, an offshoot of the
San Andreas system at the southern end of the San Joaquin Valley. It was felt
over a 160,000-square-mile area. Underground oil production facilities were
damaged on the ranch. There were twelve deaths elsewhere.

The ranch headquarters, located on the valley floor, was moved a few
miles up the Grapevine (the name of the steep grade on the north side of
the pass) to the point where the San Emigdio Mountains meet the
Tehachapi Mountains. The corporate headquarters buildings are now much
closer to the intersection of the Garlock and San Andreas Faults.

I asked the attorney about the move from the seismic frying pan into the
proverbial fire. "Interesting," he said. "It could have been instigated by
someone's wife who wanted to get out of the hot weather."

LIKE A BADGER

The remnants of Fort Tejon are preserved in a state historic park across
the interstate from the Tejon Ranch headquarters. It was this place that
gave its name to the strongest earthquake recorded during historic times in
California, a temblor of magnitude 8 dimensions that exceeded the intensity of the 1906 San Francisco quake.

When the Anglos took over the state, there were subtle references to the
past scattered about in the form of names supplied by the Spanish and

Mexicans. The Temblor Range was nearby, as were all the landscape features along the fault line named after San Emigdio.

Tejón means "badger" in Spanish, and such a ferocious adversary had confronted the holder of the first land grant at the mouth of the canyon, named Cañada des Uvas, meaning Grapevine Canyon. The location of the fort four miles north of the pass on the west side of Grapevine Creek was idyllic. The fort was an oasis at the 4,200-foot level of the pass that lay between two dusty plains.

Large oak trees, some eight feet in diameter, spread abundant shade over the deep grass, and the small brook ran all year. The creek was lined with maples festooned with grapevines. Wildlife that could be hunted for food was abundant, as were grizzly bears. An early traveler had carved the following notice in an oak tree: "Peter le Beck, killed by a Bear, Oct. 17, 1837."

Fort Tejon was the first military post constructed in the interior of California. The dragoons arrived in 1854, and the fort was completed the following year. It lay fifteen miles south of the Tejon Indian Reservation. The mission of the soldiers was to contain the friendly Indians, repel hostile ones, chase *bandidos*, and guard the principal route to northern California and the gold fields.

The fort's layout was the standard rectangular plan. Thirteen adobe structures with two-foot-thick walls faced the sloping parade ground. The only unusual features were the camels that had been imported on an experimental basis to serve as beasts of burden. The camels, horses, and mules did not mix. They grazed fitfully in their separate enclosures during the early-morning hours of January 9, 1857.

A foreshock rippled across the parade ground at about 6 A.M., but little was thought of it. Indeed, most of central California had experienced foreshocks starting late the previous night and continuing through sunrise. The main shock came at 8:24 A.M. It was felt from San Diego to north of San Francisco and Sacramento, east to Las Vegas, and southeast as far as the mouth of the Colorado River.

For a distance of 180 miles, from the Cholame Valley to Cajon Pass, the earth slipped a maximum of thirty feet along the San Andreas Fault. The shaking lasted an unusually long time—one to three minutes. Two deaths were recorded in lightly populated southern California.

For the soldiers at Fort Tejon the shocks came as a complete surprise. The fort's commanding officer, Lieutenant Colonel B. L. Beall, was in bed at the time. He was awakened, in the words of the quartermaster's deputy, "by the most terrific shock imaginable, tearing the Officer's quarters to pieces, severely damaging the Hospital, and laying flat with the ground the gable ends of nearly all the buildings erected."

From the second floor of the commanding officer's house at the head of the parade ground, Beall made his way outside as the bricks from the chimneys and plaster fell about him. The quartermaster informed the *Los Angeles Star*: "It is a miracle that no lives were lost, for which mercy we are indebted to the protecting influences of an all wise Providence." Most of the thick-walled adobe structures in the fort were damaged, and the troops moved outside in the cold weather as aftershocks continued to shake the buildings.

Nature was running amok; fear gripped the troops. Two days later the quartermaster noted: "It is very evident that a powerful volcanic eruption is in progress a few miles to the southward of the garrison. You can well imagine the alarm constantly existing in the minds of every person, caused by the frequency of these frightful shocks."

In the same general direction, a minister traveling through the area said that before the earthquake "a tremendous noise" had burst from a mountain. Others reported a "peculiar rushing or rumbling noise" that preceded the temblor and "noises somewhat resembling distant thunder" that accompanied the many aftershocks.

The manager of the fort's sawmill said he heard the same noises "and looking towards the mountain, he saw issue from its topmost peak, a mass of rock and earth, which was forced high into the air—this was unaccompanied by smoke or fire."

A petroleum seep ignited on the south side of the San Gabriel Mountains and could be seen burning at night from the San Fernando Valley. Other seeps flared along the coast. The nighttime skies took on the appearance of a Walpurgisnacht. Such flaming seeps, the clouds of thick dust, and the general turbulence and confusion contributed to the illusion of volcanoes throughout the region.

Fourteen miles to the west of the fort, at Mill Potrero, where logs were sawed for construction of the fort, huge ponderosa pine trees crashed to the ground. The mill workers lost their footing and fled in panic to the damaged fort.

Six miles to the east of Fort Tejon, at Reed's Rancho, near present-day Gorman, either a beam or a wall—the accounts vary—fell on the head of the wife of a vaquero and killed her. An old man walking toward the church in Los Angeles's plaza was the second casualty, possibly the victim of a heart attack. Neither person rated a name in contemporary accounts.

A rent in the surface of the earth left a narrow ridge of disturbed ground that was clearly visible for twenty-five miles along the road from Gorman to Elizabeth Lake. A traveler related: "Its course was in a straight direction, across valleys, through lakes, and over hills, without regard to inequality or condition of surface."

As with other quakes, there were unsubstantiated tales of objects being swallowed by the momentary opening in the ground. The *Star* cited "the great danger we were all subject to, in being engulfed by one of those visitations." There were stories of a man who managed to extricate himself from the fault and of the disappearance of a miner's blanket and rifle, a grove of oak trees, a cow, and a woman and her cabin.

The epicenter, Kerry Sieh estimated, was not at Fort Tejon. Rather, the slip began one hundred miles to the northwest at approximately the intersection of Highways 41 and 46 in the Cholame Valley.*

The tear in the fault then cascaded southeastward toward Los Angeles. It was diverted thirty degrees to the east by the Big Bend. The bend in the fault is one of two such curvatures, the second being to the east of San Bernardino at San Gorgonio Pass, where the fracture resumes a more southerly course. A number of faults join the San Andreas at these two creaky joints, also known as asperities.

Fifty years after the temblor an investigator for the 1906 earthquake commission passed through the region and remarked:

> The people living along the Rift for 150 miles southeastward from
> the Cholame Valley tell wonderful stories of openings made in the earth

---

* Ninety-eight years later actor James Dean was killed in an auto accident at this same intersection.

by the earthquake of 1857. The first settler in the Cholame Valley was erecting his cabin at that time, and it was shaken down. The surface was changed and springs broke out where there had been none before.

In 1916, Maria Ocarpia, a Salinan Indian, recounted:

> When I was a child there was an earthquake; the earth shook and the ground cracked in Cholame. We were frightened and thought that the end of the world had come. It was many years ago. The fish came out of the ground; it was a great earthquake. The animals were frightened at the water from the earthquake. The oak trees bent to the earth and the people were frightened and fell on their faces and prayed.

Just to the south of the epicenter, on the desolate Carrizo Plain, where the fault can be seen most clearly on land, the greatest stream offsets were measured. Three cowboys were herding cattle on the plain at the time of the convulsion. The cattle stampeded. The men saw dust rise where the fault ran along the foot of the Temblor Range and thought, at first, that it was from the stampeding cattle. The cowboys chased the cattle and came across what had been an O-shaped corral astride the fault. It had been bent into an S shape, a graphic reminder of the elastic nature of the earth's surface.

The Crocheron farm, located near San Bernardino on the banks of Lytle Creek, was at the other end of the fault line that was displaced by the 1857 quake. Later in life Augusta J. Crocheron wrote a vivid description of the earthquake, one which captures the essence of such an event, if encountered outdoors:

> Can anyone who has ever experienced an earthquake, overcome a dread of its recurrence; or mistake the signs that are usually premonitors of its coming? One pleasant morning I was searching through garden paths for roses for the breakfast table, when the air seemed to hold still, not a breath stirring. I heard a far off smothered, rumbling sound, that I scarcely noticed, for I thought I was growing dizzy, and not understanding why I should feel so, I started for the house. As I stepped across a narrow stream, the opposite bank seemed first to recede from me, then instantly to heave upward against my feet. As this threw me from my equilibrium, the water emptied out on either bank, and hearing an Indian's voice in loud supplication, I turned and saw our Lothario on his

knees, the ground rising and falling in billows around him. At the same instant I saw my parents and sisters clinging to large trees, whose branches lashed the ground, birds flew irregularly through the air shrieking, horses screamed, cattle fell bellowing on their knees, even the domestic feathered tribe were filled with consternation. Voices of all creatures, the rattling of household articles, the cracking of boards, the falling of bricks, the splashing of water in wells, the falling of rocks in the mountains and the artillery-like voice of the earthquake, and even that awful sound of the earth rending open—all at once, all within a few seconds, with the skies darkened and the earth rising and falling beneath the feet—were the work of an earthquake. It passed—we rejoined each other, thankful that life was spared, and looking around with trembling, upon the scene, where utmost terror had reigned. Said father, it is scarcely time to congratulate ourselves, another shock may occur in half an hour. In suspense we waited, and it came. Then the skies cleared, the air moved with cool, swift wings, the stream ran clear, and the earthquake's spell had passed. When we ventured to walk around at a little distance from the house, we found, about twenty rods away, a rift in the solid ground, a foot wide, a hundred feet long, and so dark and deep, we feared even to measure it.

The noise of the upheaval was particularly noticeable in the San Bernardino area. Other residents referred to it as "a rumbling" sound, "a terrible report," and a "loud report similar to the discharge of Cannon" in the nearby mountains. Fifty miles to the northeast, in the Mojave Desert, "a peculiar harsh, grating noise" was heard by a surveyor at his camp on the banks of the Mojave River.

Complaints of dizziness, nausea, and vomiting were prevalent throughout southern California, perhaps because the long time-span of the quake and its swaying, described as "slow and gradual," resembled the rhythm of an ocean voyage.

Forty miles directly south of the fault and 150 miles southeast of the epicenter, the quake was less noticeable but still quite frightening in Los Angeles. The frontier town of some four thousand inhabitants, who lived in one-story adobe structures, was rife with public drunkenness and violence.

The *Star* reported that when the earthquake struck, "The people fled into the streets; many could not stand and in terror fell to their knees and cried out, 'Lord have mercy.' " *El Clamor Público*, the newspaper that served the Hispanic population, reported: "It seemed as though the Earth, tired of suffering our sins, was shaking herself free of us as birds shake off what disturbs their feathers."

Many people were temporarily sick in Los Angeles, but few were injured. The inhabitants were saved because they had light roofs coated with asphalt from the La Brea tar pits rather than the usual heavy tiles. The heavy Spanish tile roof of the old church at the Ventura Mission collapsed, but fortunately there was no service at the time. A member of a Coast and Geodetic Survey team inspected the wreckage. He estimated that the walls of the church had supported a load of a half million pounds.

In nearby Santa Barbara, where the damage was light, the newspaper editor was more concerned with the paucity of the English language when it came to describing such events. "Shock," the most expressive word available at the time, was a poor substitute for the more graphic "temblor" derived from the Spanish, he noted.

The aftermath of the quake set a precedent that was followed for years to come in California. The *Los Angeles Star*, the newspaper that served the Anglo population, was quick to reassure its readers. Its self-serving attitude toward natural disasters came to typify the reaction toward earthquakes. Earthquakes were bad for business, and newspapers were a business and vigorous cheerleaders for a prosperous economy. The editor wrote:

> We have taken some trouble to inform ourselves on the subject generally, and we have come to the conclusion that no great danger need be apprehended from earthquakes in this country—judging of the future by the past, and we suppose the laws of nature will not be materially departed from, either to punish or reward the new race which now possesses the land.

By 1857 the many newspapers in post–gold rush California had replaced church records as the seismographs of their time. At the time there were few geologists with formal training in the state, and most of them were concerned with mining matters. The newspapers collected and made public the seismic observations of others and came to their own conclusions, some erroneous and some quite astute.

Just by observing and using common sense, journalists were the first to suggest the presence of a continuous fault line and note the problems with various building materials. But they did not follow up on their suppositions and were soon pursuing other stories. Later they deferred to formal learning.

One week after the earthquake the editor of the *Los Angeles Star* voiced concern for the safety of San Francisco because he could see "the line of demarcation" extending in that direction. It was a prescient comment.

The first extensive description of a continuous fault and the type of movement that was typical of the San Andreas was written by a newspaper editor. Stephen Barton, editor of the *Visalia Iron Age* in the San Joaquin Valley, published a weekly series of articles on the "History of Tulare County" that ended in December of 1876. In one article he referred to "the line of disturbing force" and continued:

> This line was marked by a fracture of the earth's surface, continuing in one uniform direction for a distance of two hundred miles. The fracture presented an appearance as if the earth had been bisected, and the parts had slipped upon each other. Sometimes the earth on one side would be several feet higher, the highest presenting a perpendicular wall of earth or rocks. In some places the sliding movement seems to have been horizontal, one side of the fracture indicating a movement to the northwest, the other to the southeast. The fracture pursued its course over hill and hollow, and sometimes this sliding movement would give to the points of the hills and to gulch channels a disjointed appearance.

WALLACE CREEK

The Carrizo Plain lies fifty-five miles northwest of Tejon Pass and is just south of the epicenter of the 1857 quake. No section of the fault line is more remote, barren, forbidding, or representative of a tectonic landscape.

It is well worth a visit, but the casual traveler needs to take precautions. Arrive with as full a tank of gas as possible and some water and food. Tourist facilities are nearly nonexistent. There are a few ranches, and a visitor center run by the federal Bureau of Land Management. (For timely information call 805-475-2131.) Avoid the hot summer months when the landscape is baked to the consistency of burnt bread. Shun the rainy weather that turns the dirt roads to mocha pudding.

The plain is a high desert, separate from the remainder of the state except for the umbilical cord of the fault. The vegetation is sparse; there are few trees. Rainfall is slight and temperatures can soar over one hundred degrees. The sun and omnipresent sky are the dominant exterior forces. I camped for a number of days on the fault line but felt no interior movement. Others have.

The Chumash Indians believed they lived on a big island that was the center of the universe. Two giant serpents held up the island, and when one moved there was an earthquake. A serpent is painted on an outcrop of rock on the Carrizo Plain.

The Spaniards arrived next, and the following miracle was recorded:

> The priest who had accompanied the explorers prayed earnestly that water might be found, and soon; and shortly thereafter a great earthquake shook the mountains, ripping a deep gash through the rock formations, and from this break in the earth's surface there gushed forth a stream of pure cool water, released from its underground channel.

The spot was called El Campo de Los Temblores, and from it the Temblor mountains to the east got their name. To the west, the mountains were christened the Calientes, meaning "hot."

Beginning in 1906 and proceeding to the present time, numerous geologists rode across the plain either on horseback or in motor vehicles. They were attracted by the visible cleavage of the fault and the large stream offsets. Here was a chance to make their mark. No other single section of the San Andreas Fault has been studied more extensively.

Most of the studies have centered in or around Wallace Creek, reachable most easily from Highway 58 on the east side of the plain. It was here that a stream bed was deflected more than four hundred feet over thousands of years. The land moved thirty-one feet in 1857, believed to be the greatest amount of offset generated by any single earthquake during historic times in the contiguous United States.

From the paved highway I took Seven Mile Road west a short distance and turned left, or south, onto Elkhorn Road. After I passed underneath the high-voltage electrical transmission lines, there was a sign stating "Wallace Creek" and a fenced parking lot. I walked north along the fence to the misnamed creek. It is dry far more often than it is wet and thus qualifies more accurately as an arroyo.

The straight-edged bluff was the fault line. I looked at the ninety-degree angle where the creek met the fault, its sharp jog and clearly incised 430-foot offset along the rift line, and its abrupt column-left exit and march down the gentle slope. The angular path resembled a Z with its middle section straightened. It was clear this was no ordinary watercourse.

The dry stream bed was named after Robert E. Wallace, a USGS geologist who edited a comprehensive study of the San Andreas Fault. Others had been there before him, but Wallace was the first to write about stream offsets and have his findings published.

In 1968 Wallace reported that 130 stream channels had been offset along seventy miles of the fault in this area, some by as much as 1,200 feet over many years. Some stream beds were offset 30 feet by the 1857 earthquake alone. Wallace's figures hinted at the number of large temblors that must have rippled over the plain's surface in relatively recent geologic times. "Even if each event produced as large an offset as 30 feet, forty such events would be required to produce the cumulative offset of 1,200 feet," he wrote.

By 1975, Wallace had found total stream offsets greater than 3,280 feet. He estimated that large earthquakes might occur on an average of every 700 years; or to put it another way—there was a one in seven hundred chance of such a temblor. He also cited the possibility of irregular clustering.

Kerry Sieh had looked at Wallace Creek before going to work at Pallett Creek. He was aware of Wallace's findings. Sieh returned to Wallace Creek after obtaining his degree and beginning a teaching and research career at Caltech. He dug trenches and published the results in 1984.

Certainty was in the air at the time, prediction seemed doable in that decade, and paleoseismology appeared to have many of the answers.

Sieh verified Wallace's findings and refined the numbers. Small gullies nearby indicated that the land slipped 31 feet in 1857. The 430-foot offset occurred over a period of 3,700 years. There were larger offsets over greater spans of time. He predicted there would be a large earthquake by the end of the century just to the northwest of Wallace Creek, but a seismic event was not likely to the southeast for at least another 100 years, he thought.

Sieh and Wallace teamed up in 1987 and wrote an article stating: "Rarely are tectonic landforms as well expressed and as well dated as they are at Wallace Creek. Along this 1.5 mile length of the San Andreas fault are examples of most of the classic geomorphic features of strike-slip faults, including offset and beheaded channels, shutter ridges, and sags."

They corrected a common misconception by pointing out that the total plate motion took place on the much wider system of parallel faults. "Contrary to popular belief," they wrote, "the plate boundary cannot be straddled by standing astride the San Andreas Fault."

A younger generation began making their mark on the plain in the 1990s. Among them was Lisa B. Grant, who had studied under Sieh. They co-authored two articles.

It was a different era. Predictions elsewhere had failed. Critics pointed out discrepancies in paleoseismic findings. Uncertainty was rife. Complexity had reared its ugly head. A textbook published in 1996, which singled out Sieh's contribution to popularizing the concept of historic events, nevertheless cautioned: "On a fault that has slipped for millions of years, even a 3,000-year record covers only a fraction of a percent of the history of the fault."

More trenches were dug at two sites just to the southeast of Wallace Creek. The lengths of offsets were further refined, this time in fractions of a foot.

Grant's and Sieh's 1993 and 1994 papers cited "uncertainties in the interpretation of offset geomorphic features," "sparse" data, "primitive" models of fault behavior, and "irregular recurrence intervals for prehistoric earthquakes on the San Andreas Fault." From a concept of "nearly periodic large earthquakes separated by unusually long recurrence intervals" in the previous decade, Sieh moved to a notion of "irregular earthquake occurrence." The two researchers even hinted at the "chaotic, rather than time-predictable or characteristic" nature of seismic events.

They determined there had been five large earthquakes on the Carrizo Plain since the start of the thirteenth century. The Fort Tejon quake was not the greatest; the penultimate event rocked the region in the fifteenth century. There was little correlation between the paleoseismic findings from trenches on the Carrizo Plain, at Pallett Creek, and at Wrightwood.

Thus, it was unknown what Los Angeles could expect from the San Andreas Fault in the future.

My visit to Wallace Creek spawned its own little mystery.

I walked the fault line along the creek bed, which resembled a military trench in its exact alignment. Only the banks and the sky were visible from its depths. It was a perfect hiding place.

There was more grass, albeit skimpy, at one spot; and sprinkled about were bits of charcoal. A man-made fire had fertilized the pebbly stream bed.

Among the small chunks of burnt wood, I found whitened bones that appeared to be from an avian creature, a rabbit, and a cow or lamb. Mixed in was the remnant of a red 1995 registration sticker from a California license plate. A flat piece of burnt material, perhaps plastic, was calcified and pierced by a rivet.

To figure out what all of this meant, I thought, would be as difficult a task as determining the past seismic history of the Carrizo Plain in order to ascertain what would occur in the future. It would take a great deal of imagination to make a coherent story from these disparate items.

5 HAYWARD
(1868)

MARK TWAIN AND THE FIRST GREAT SAN FRANCISCO QUAKE

If there ever was a person in dire straits who could find something humorous to say about an earthquake, that person was Mark Twain. He had just been fired, or rather asked to resign, from his job as the sole newspaper reporter for the *Morning Call* in San Francisco. Twain was in debt, thought himself a failure as a writer, and was contemplating suicide. Poison or a pistol were mentioned in a letter. He had submitted a story about a jumping frog in the fall of 1865, but it had not yet been published.

The sunny October morning was unusually peaceful when Twain and most of the other residents of San Francisco, veterans of numerous fires but virgins when it came to damaging earthquakes, encountered their first real temblor. The *Daily Alta California*, the newspaper with the largest circulation in the state, noted: "Our city and state had a great sensation yesterday in the way of an earthquake. It was undoubtedly the most severe shock felt in San Francisco since the American conquest." It was quickly dubbed "the Great San Francisco Earthquake," as were successively larger shocks in 1868 and 1906.

Twain portrayed the October 8, 1865, quake, centered in the Santa Cruz Mountains and later ranked "strong" by the USGS, thusly:

As I turned the corner, around a frame house, there was a great rattle and jar, and it occurred to me that here was an item!—no doubt a fight in that house. Before I could turn and seek the door, there came a really terrific shock; the ground seemed to roll under me in waves, interrupted by a violent jiggling up and down, and there was a heavy grinding noise as of brick houses rubbing together. I fell up against the frame house and hurt my elbow. I knew what it was, now, and from mere reportorial instinct, nothing else, took out my watch and noted the time of day; at that moment a third and still severer shock came, and as I reeled about on the pavement trying to keep my footing, I saw a sight! The entire front of a tall four-story brick building in Third street sprung outward like a door and fell sprawling across the street, raising dust like a great volume of smoke! And here came the buggy—overboard went the man, and in less time than I can tell it the vehicle was distributed in small fragments along three hundred yards of street. One could have fancied that somebody had fired a charge of chair-rounds and rags down the thoroughfare. The street car had stopped, the horses were rearing and plunging, the passengers were pouring out at both ends, and one fat man had crashed half way through a glass window on one side of the car, got wedged fast and was squirming and screaming like an impaled madman. Every door, of every house, as far as the eye could reach, was vomiting a stream of human beings; and almost before one could execute a wink and begin another, there was a massed multitude of people stretching in endless procession down every street my position commanded. Never was solemn solitude turned into teeming life quicker.

The earthquake destroyed San Francisco's city hall—a structure also destined for damage or destruction in 1868, 1906, and 1989—and several other poorly built structures on "made land," meaning land that had been created by filling in the bay. Such land was particularly susceptible to liquefaction, a process whereby wet soil, when violently shaken, is transformed to the consistency of thick soup. There was severe damage as far north as Petaluma. Cracks in the ground were reported along what would become known as the San Andreas Fault.

Twain was on his way to literary celebrity and a life spent elsewhere after "The Celebrated Jumping Frog of Calaveras County" was published in November. He never forgot the only time he was fired. Shortly after the 1906 earthquake, Twain saw a newspaper photograph of the wreckage of the *Call* building and wrote: "I had never lost my confidence in Providence during all that time. It was put off longer than I was expecting, but it was now comprehensive and satisfactory enough to make up for that."

BIRTH OF THE POLICY OF ASSUMED INDIFFERENCE

Three years later the second "great" San Francisco earthquake struck across the bay in Hayward, then called Haywards. The 1868 shock had all the principal elements of what was to come in 1906, but unfortunately nothing was learned from the earlier experience. It was as if the Bay Area's famous fog acted as an amnesiac.

The first warning was a deep, rumbling noise that seemed to come from the direction of the ocean shortly before 8 A.M. on October 21, 1868. Then a forty-five-second shock followed. That was three times as long as the 1989 Loma Prieta quake and can seem like an eternity for anyone who experiences it.

George A. Goodell felt it and recalled years later:

> Since October 5, 1862, I have lived in Haywards, Alameda County, and I well remember the earthquake of October, 1868. Being lame and having used a cane from childhood, I had never walked without it until that morning. I was working in my shop at the time. On feeling the terrible shock, and on the impulse of the moment, I managed to get out of the building and into the street, some 18 feet distant, but on recovering from my fright I found I had left my cane in the shop. I managed to get back into the building, got my cane, and started for my house only a few yards away. The house had been thrown from its foundations, the chimney had been torn from the roof, and the porch had been wrencht away. Dishes were broken and everything was in confusion. I discovered that most of the houses were in the same condition as my own—thrown from their foundations, with chimneys down, porches knocked sideways, etc. All the while the ground was shaking and continued to shake for days and even weeks; but each shock was lighter than the last.

Goodell continued, "At San Leandro the earthquake destroyed the brick courthouse, which was then located there. A Mr. Joslyn was killed in attempting to escape from the building. Many buildings were much damaged in that town as well as in Haywards. The earthquake was the direct cause of the death of 2 persons in Haywards."

The damage was greatest in the East Bay, since the quake was centered on the Hayward Fault, which ruptured along a thirty-mile front. There also may have been an earthquake on the same fault, a major branch of the San Andreas, in 1836; but accounts were sketchy. (Recent research places this quake near Monterey.) For the first time, as far as was known, an earthquake was accompanied by deaths in northern California—a total of thirty in 1868.

In retrospect, it can be seen that a pattern was emerging. The parts of San Francisco that suffered the greatest quake damage in 1865, 1868, 1906, and 1989—namely the "made lands"—were equidistant from the San Andreas and Hayward Faults. The Hayward quake had a moderate magnitude roughly equivalent to the 1989 Loma Prieta temblor on the San Andreas Fault. All quakes except the 1865 temblor were accompanied by fires.

Looking back at 1868 from the perspective of the present decade, historian Charles Wollenberg wrote, "Coming so soon after a massive southern California earthquake in 1857 and a far less damaging Bay Area tremor in 1865, the 1868 quake clearly established that California was 'earthquake country' and that highly destructive seismic upheavals were a fact of life."

In 1868 the population of San Francisco was 150,000, which was more than one-fourth of the state's inhabitants, and the city was growing fast. The boomer mentality prevailed. The business interests and the newspapers wanted to keep it that way, despite the serious flaws in the landscape.

Humans, not the destructive power of earthquakes, were blamed, since to admit to the presence of uncontrollable seismic forces was to discourage the investment of capital and continued migration to this city by the gorgeous bay. Build better and more expensively was the rallying cry, as it had been in 1865.

The *Daily Alta California* reported that "crowds of idlers" on the day of the quake gave way to "the hum of busy industry" on the next day. "There was nothing of despair, discouragement, or even doubt of the future to be seen on the countenances of our citizens," noted the paper, "but everywhere a fixed determination to repair losses and do better work than before." There was little discussion in the press about the origin of the quake, science being in its infancy on the West Coast.

To make sure the message got out to the financial markets in New York and London, the chamber of commerce appointed a committee of the wealthiest merchants in town to send an upbeat telegram. It ended with these words: "No damage to well-constructed buildings. Total loss on property will not exceed $300,000." Actually, estimates reached five times that amount.

Then there was the matter of "the report."

There was a firm belief within the California scientific community for more than a century that a report on the 1868 earthquake had been suppressed. The suspicion took on the aspect of paranoia among scientists because of the undeviating Panglossian attitude of the business community toward earthquakes, and the opposition to efforts to understand and explain them. The rumor was codified in the report of the state commission appointed by the governor to investigate the 1906 quake.

The commission's report stated:

> Shortly after the earthquake of 1868 a committee of scientific men undertook the collection of data concerning the effects of the shock, but their report was never published nor can any trace of it be found, altho some of the members of the committee are still living. It is stated that the report was supprest by the authorities, thru fear that its publication would damage the reputation of the city.

The commission based its conclusion on the recollection of one of its members, George Davidson, who was also a member of a chamber of commerce subcommittee that looked into the scientific aspects of the 1868 quake. Davidson stood at the very apex of the state's scientific establishment. For many years he served as president of the California Academy of Sciences and in 1906 was elected the first president of the Seismological Society of America.

In a newspaper interview following a moderate quake in San Francisco in 1898 Davidson said:

> A scientific committee was appointed to investigate the affair. The merchants wanted to know all about it in order to determine what should be done to make them more secure in their houses. I was a member of that committee, and I remember that the result of our inves-

tigation was so startling that we never published our report. It was thought that to do so would frighten people who intended to come here to settle.

Two years after the 1906 quake and forty years after the incident—if, indeed, it was that—Davidson wrote:

> The Sub Committee made a detailed report of what had been examined, estimated the damage done in San Francisco at $1,500,000, and made certain recommendations upon the necessity for good foundations, and other matters of interest. The report was carefully prepared but Mr. Gordon [a businessman who headed the main committee] declared that it would ruin the commercial prospects of San Francisco to admit the large amount of damage and the cost thereof, and declared he would never publish it. I know of no one who had taken a copy of the report.

There is another paper trail that suggests an alternate fate for "the report." The chamber of commerce's Joint Committee on Earthquake Topics lacked expertise, quickly lost interest in the subject, failed to raise money for the study, and it died a premature but natural death.

A secretary to the committee, an Englishman by the name of Thomas Rowlandson who had some expertise in seismic matters, cited the "desultory conversational character" of the committee's meetings. He added, rather sarcastically, that publication of a report was "awaiting, probably, the time when these amateur architects, chemists, geologists and general scientists shall have, to their own satisfaction, mastered *the alphabets* of the particular sciences on which they have undertaken to treat."

Rowlandson, a fellow of the Geological Society of London, printed his own ninety-six-page report but doubted that there was sufficient interest to repay the cost of publication from sales. "Unsuitable" mortar used in brick structures and the lack of reinforcement were the main reasons for the collapse of buildings, he concluded.*

There was mention of yet another report in 1906. At the time of the earthquake the *Mining and Scientific Press* noted:

---

* There is a slight chance that Rowlandson's report was "the report." It is available in the Huntington Library in Pasadena.

There is room for comment here on the short memories of humankind. Here was an authority—E. S. Holden—who reported on the last serious earthquake forty years ago, and emphasized the danger of building 'on the made land between Montgomery street and the Bay.' Nevertheless, during the years since that report was published, people have continued to build there in disregard of the evidence that it was dangerous to do so without taking proper precautions.

Holden was a former president of the University of California and director of the university's Lick Observatory. He operated the state's first seismographs and compiled a catalog of California earthquakes.

Whether suppressed, nonexistent, or simply disregarded at the time, it made little difference. No report on an earthquake, either then or now, has affected public policy. Reports on this subject—even very good ones, as we shall see—are not popular.

Josiah Dwight Whitney, whose extensive geological survey of the state ground to a halt in 1868 because of lack of funds, described the attitude in California toward earthquakes:

> The prevailing tone in that region, at present, is that of assumed indifference to the dangers of earthquake calamities—the author of a voluminous work on California, recently published in San Francisco, even going so far as to speak of earthquakes as "harmless disturbances." But earthquakes are not to be bluffed off. They will come, and will do a great deal of damage.*

Whitney's prophetic words were echoed and enlarged upon forty years later by Grove Karl Gilbert, the one scientist who did more than any other to advance the knowledge of earthquakes in this country. Gilbert was a highly respected geologist who worked for the USGS. He commented on California's earthquake policies and the missing report in a 1909 presidential address to the American Association of Geographers. Gilbert said:

---

* When Whitney, an easterner, first arrived in San Francisco in 1860 he wrote his brother: "Next to the earthquakes what shocks the stranger most is the scale of prices in this region." Little has changed during the intervening years.

The policy of assumed indifference, which is probably not shared by any other earthquake district in the world, has continued to the present time and is accompanied by a policy of concealment. It is feared that if the ground of California has a reputation for instability, the flow of immigration will be checked, capital will go elsewhere, and business activity will be impaired. Under the influence of this fear, a scientific report on the earthquake of 1868 was suppressed.

THE HAYWARD FAULT AND THE POLICY OF ASSUMED INDIFFERENCE TODAY

The policy of assumed indifference is alive and well in California and is best illustrated by a stroll along the Hayward Fault, a branch of the San Andreas Fault system.

Of the two parallel faults that bracket San Francisco Bay and presently hold millions of people in their potentially explosive embraces, the Hayward is given the best chance of rupturing in the near future. The fault extends fifty miles along the foothills of the East Bay from San Pablo Bay on the north to east of downtown San Jose on the south. Under the names of the Rodgers Creek, Healdsburg, Maacama, and Garberville Faults, the Hayward Fault actually extends all the way to Humboldt County.

No other segment of the San Andreas Fault system passes through such a dense urban population. The fault bisects private residences, businesses, and industrial complexes in the cities of Richmond, Berkeley, Oakland, San Leandro, Hayward, and Fremont. On the fault line are such public facilities as schools, colleges, football stadiums, freeways, a rapid transit system, hospitals, a civic center, the aqueduct that transports water from the Sierra Nevada to San Francisco, and the Berkeley campus of the University of California, where seismology was first taught in this country.

The prestigious university, I thought, would be an ideal place to learn how a large institution dealt with the complexities of seismic movement.

My first stop was the seismograph station on the fifth floor of McCone Hall, only a few hundred feet west of the fault. I was quite familiar with the building, having walked past it four times a week when I taught at Cal for a number of years. I have also used the Earth Sciences Library on the second floor to research a number of books, including this one.

The oldest continuously operating seismograph facility in the Western Hemisphere, with twelve recording stations scattered throughout northern California, the main Berkeley seismograph station has moved a number of times on the campus that overlooks San Francisco Bay. When I visited it, the station was located down a long, dark corridor on the top floor of McCone Hall, also known as the Earth Sciences Building.

There were no rotating drums and pens making squiggly lines on graph paper. The operation was digital. The Berkeley stations searched for the larger temblors, while the U.S. Geological Survey network, based across the bay in Menlo Park, was keyed toward microquakes.

On the day I stopped by, the rooms on the southeast corner of the building were filled with students at computer workstations. Earthquake and epicenter maps were tacked on walls, and technical journals and coffeemakers were scattered about. Lying on a desk was a "Primary Response Notebook" with nine steps—the fourth being "process event," and the last, "prepare for media."

Graduate students and a research associate called up vast amounts of data from two Sony mass-storage devices, nicknamed "jukeboxes." Each contained six hundred gigabytes of information gathered from stations in northern California and instruments operated by institutions located elsewhere. The Berkeley archive contains nearly ninety years of continuous records, a length of time greater than any other observatory in the country. Since 1910 more than 150,000 quakes have been recorded.

I watched the sedentary researchers crunch numbers and listened to their arcane jargon, which resembled medieval numerology to my uninitiated ears. I was reminded of a warning voiced more than fifty years ago by Perry Byerly, a former director of the observatory. He wrote: "A complete description of an earthquake can never be made from the study of the records of seismographs. In this day of the extensive use of instruments there is grave danger that field observation will be neglected."

I walked over to the corner and watched Rick McKenzie, a staff research associate, massage the data from an earthquake near San Jose a few days previously. There were erratic EKG-type lines in the zoom window of McKenzie's computer screen, and vast amounts of data from the PC-based broadband digital seismographs scrolled past in multiple columns.

. . . .

I was not aware, until I probed the documents of a court case, that the seismological observatory probably would not be operable if there was a moderate to severe earthquake on the Hayward Fault.

Built in 1961, McCone Hall subsequently ranked within the top 6 percent of those structures needing seismic upgrading in a survey of 3,500 state buildings in California. In 1978 McCone was given a "poor" seismic rating. A 1990 seismic improvement study of the Earth Sciences Building determined that the south side, where the seismological station was located, rested on soft ground and the exterior wall lacked the necessary tensile strength.

In fact, there was a weakness in the south wall between the foundation and the second story. The concrete columns, the basic support, lacked ductility. The first floor thus became a "soft story," and in a major quake the building would collapse. Additionally, a large concrete overhang on the fifth floor would come crashing to the ground, squashing anyone underneath it. In the bureaucratic language of the 1990 study, the building's occupants faced "appreciable life hazards."

University officials had some money left from a larger retrofitting project, so they decided to fix McCone. An Oakland architectural firm was hired. The resulting plan passed peer review and was approved by the university's Seismic Review Committee at a meeting missed by one of its members, Stephen A. Mahin.

Mahin, a professor of structural engineering, was on sabbatical leave. When he returned briefly to the campus in November of 1993, he walked by McCone and stopped to examine the construction. Mahin had good reasons for being interested in what was going on. The type of concrete wall being installed was one of his specialties, and the original structural engineer who designed the building was Byron Nishkian. Mahin was the Byron L. and Elvira E. Nishkian Professor of Structural Engineering.

What Mahin saw disturbed him. He fired off a memo to the assistant chancellor in charge of planning, design, and construction. It stated, "I found several aspects of the detailing being utilized in the retrofit of McCone Hall to be quite unusual for structures located in seismic regions in general, let alone those located within a half mile of a major earthquake fault." What disturbed Mahin was that certain construction details might not meet the requirements of the Uniform Building Code. Mahin then departed for China, leaving university officials to deal with the problem.

They reacted immediately. Meetings were called. Faxes, letters, and phone calls flew back and forth. There were additional peer reviews by consulting engineers. One structural engineer supported the architect, citing the architectural and budgetary constraints the university had placed on the project. She wrote, "The engineering design, by trying to simultaneously satisfy all the requirements, becomes as much an art as a science."

Another engineer with a longer university affiliation found serious flaws in the design. The university sided with the second engineer, stopped work on the project, fired the architect, and patched up the exterior of the building without completing the job. Lawyers became involved, the university demanded a payment of $1.32 million from the architect, and a lawsuit was filed.

After the southern California earthquake of January 1994, near panic struck the occupants of the building. The chairs of the Departments of Geography, Geology, and Integrative Biology (formerly paleontology) and professors in the various departments sent angry letters to associate deans, deans, vice chancellors, and the chancellor of the university.

One chairman cited "this whole bizarre performance" and "the manifest culpability of university employees and contractors in the present debacle." Requesting that "in these extraordinary circumstances" the dean take responsibility for the safety of the faculty, staff, graduate students, and undergraduates, he warned:

> The administration should be in no doubt that if death and injury results from a seismic event before a retrofit is completed the claims for punitive damages against the Regents will be enormous, given the admitted incompetence of everyone concerned in the present fiasco.

Barbara A. Romanowicz, director of the seismological stations, wrote the person in charge of space management at the university—with a copy sent to the chancellor—that a moderate-sized quake would disrupt the operations of the seismographic station. She said one of the problems would be access to the building to obtain readings for the media and public.

Citing the 1994 Northridge earthquake and the recent upgrading of the station's facilities in order to participate with other institutions in a new program aimed at rapidly pinpointing such quakes, the director demanded *"that the Seismographic Station be promptly relocated in a seismically safe building on campus."* The station remained in McCone.

Chancellor Chang-Lin Tien replied with soothing assurances. To one professor he said that "some projects are more difficult to manage than others; the McCone Hall Seismic Retrofit may end up being at the top of the list." The job, redesigned by a different architect, got under way for a second time in 1997.

From McCone Hall I walked southeast and crossed Piedmont Avenue to Memorial Stadium, which straddles the Hayward Fault. The university had commissioned a geology study and knew about the rift before it built the monumental stadium in 1923.

The chairman of the stadium committee said that movement along the fault line was "due to chemical rather than tectonic forces. It does not affect the Stadium site in any way." The chairman was dean of the College of Mining and presumably knew something about geology. He was also an ardent fan, who wrote in an article celebrating the completion of the stadium that "to red-blooded men the pigskin is at times mightier than the sheepskin."

There were other sites, ones flatter and more accessible; but the university did not own them and could cut costs if a football stadium was built on land it owned in Strawberry Canyon. With the fervor of a religious crusade, the university raised one million dollars in a three-week subscription campaign to build a coliseum-type structure, one that would fit into the Athenian design concept for the campus.

A great deal of blind boosterism was involved in the campaign. University officials unabashedly beat the drums, stating of the athletic facility: "It challenges the admiration of the world, an architectural monument, an engineering accomplishment ranking with the greatest of all times." Comparisons were made with the engineering of the Panama Canal and the architecture of Athens and Rome.

William Henry Smyth, a neighbor, was what is now known as an environmentalist—or less kindly, a NIMBY (not in my backyard). Smyth objected to "sacrificing one of Nature's priceless gems to the purposes of commercialized 'sport.' " He noted that the canyon site was hemmed in on two sides by steep hills and wondered if it could be vacated quickly in an emergency.

Smyth's curiosity was piqued as he watched the workmen dig the huge foundation for the eighty-thousand-seat stadium. He wondered: what was that unusual soil, and why didn't Strawberry Creek head straight west toward the bay instead of jogging north, then west? The zigzag course of the creek was a clue to a fault offset that had previously been recognized by the university.

The inquisitive neighbor found the answers to his questions in a geologic map. Smyth wrote in a pamphlet, *The Story of the Stadium*, published the same year the facility was completed, that a fault "extends across the mouth of Strawberry Canyon and thence through the Greek Theater northward through the University Campus, and southwardly through my front garden."

He did not associate the fracture with the devastating 1868 earthquake, nor apparently did anyone else—publicly. Today construction would have come to a screeching halt. Indeed, it would never have commenced, what with all the required environmental and geologic studies. But in 1923 there was nary a blip. The stadium was completed in eleven months, just in time for seventy-three thousand fans to watch Cal defeat Stanford in the Big Game on November 24. Nirvana had been reached.

Through the years, Cal teams—not recently known for their winning ways—have played before crowds of seventy thousand some fifty times; and at least twenty games have attracted more than eighty thousand spectators. Ninety thousand crowded into the oval stadium to hear President John F. Kennedy speak on March 23, 1962.

During this time, university officials were aware of the creeping fault and the damage it caused. Water lines in nearby buildings frequently broke. The bypass culvert that carried Strawberry Creek water under the stadium cracked under the constant strain and needed frequent repairs. A rip appeared in the concrete wall of the stadium, and gradually it widened.

In 1938, the administration was considering building a women's dormitory in the fault zone and sent an employee to sound out George D. Louderback in the geology department on the seismic feasibility of the project. The subordinate's memo on the conversation with Louderback noted:

In reference to the California Memorial Stadium, the fault plane passes directly through the structure, and on account of the massive type of structure there would be actual rupture in case of movement at the time of an earthquake. At such time, if the Stadium were filled with people, the greatest loss of life would probably be from the ensuing panic.

As for the dormitory project: "Of course, I did not get any yes or no answers," reported the employee to the university comptroller. Louderback was subsequently hired as a consulting geologist on the project. His final report stated, "The reasonable expectancy is that it will be free from the risk of future fault displacements (although not free from the risk of strong earthquake shock)."

Stern Hall was subsequently built. Stern Hall and the Greek Theater, used for large public gatherings, are within the fault zone. Nearby Bowles Residence Hall sits atop the actual fissure, as does the stadium. McCone Hall is nearby.

The university does not advertise the presence of the fault nor the vulnerability of its structures.

The California Seismic Safety Commission sought to have the university post seismic warning signs on its buildings after the 1989 Loma Prieta earthquake, but the administration opposed that move. Referring to the main library as an example, university president David Gardiner said:

> Put a big sign on the Doe Library at Berkeley: "This building is seismically unsafe. Enter at your own peril." Okay. So you're a freshman student coming up to the library. What are you supposed to do about it? Some students say, "Oh, that's fine," and they walk right in. Other students say, "Well, I'm not going in there."
>
> So that's what came up in my mind when I heard this suggestion. That seems to me not to be very helpful, for what should be self-evident reasons. What is the student supposed to do about it?

Regarding the stadium, the *Golden Bears Media Guide* states:

> Now 72 seasons after its 1923 opening, the setting of Memorial Stadium remains one of the most breathtaking sights in all of college

athletics. The plush wall of pine trees in the Berkeley Hills to the east is contrasted by a panoramic view of the San Francisco Bay and three bridges to the west.

No mention of the Hayward Fault, the earthquake of 1868, the high probability of another quake on the same fault, the cracks in the stadium, the consultant's warning about locating athletic department offices under the stands, or the fact that the new press box has a support system independent of the main structure mars the hyperbolic prose of that promotional document. As for the future of the stadium, the university has studied the problem—termed "difficult"—but made no decision.

I climbed to the top row of paint-flaked wooden benches in Section KK of the aging structure. A bent, rusted steel plate was bolted to the concrete cap of the stadium in a futile effort to conceal the fissure that spreads to ground level. The fault creeps at an average of one-tenth of an inch a year. At the top, the west side of the stadium has advanced four inches beyond the east side.

I peered inside the crack and could see that the steel rebar was bent to accommodate the strain. The concrete was ripped apart, revealing the aggregate buried inside. I sat with my back to the concrete that had been warmed by the sun and thought about the playing field far below me. The force that could rend a stadium and a region seemed colossal in comparison to the tiny figures who had fleetingly dashed across the manicured green turf over the years.

I have never watched a football game in Memorial Stadium, nor do I plan to in the future, especially after I drove home and read this message from the seismograph station posted on the Internet: "The Hayward Fault runs under Berkeley Stadium and poses a severe hazard if a damaging earthquake occurs during a football game when tens of thousands of people are in the stands."

A SLOWLY EVOLVING SCIENCE

During the years following 1868, when science was rapidly advancing elsewhere, California's scientific community dozed in the somnolence of the Pacific sunshine. Through most of the nineteenth century, the science of seismology evolved in Europe—most notably in England, Italy, and Germany. By the end of the century, the Japanese had advanced the furthest.

It was the ever-curious, restless English who led the way in the West. Charles Darwin was traveling on the *Beagle*, gathering material for his great work on evolution, when a tremendous earthquake struck Concepción, Chile, on February 20, 1835. He wrote:

> A bad earthquake at once destroys our oldest associations: the earth, the very emblem of solidity, has moved beneath our feet like a thin crust over fluid;—one second of time has created in the mind a strange idea of insecurity, which hours of reflection would not have produced.

One hour and forty minutes before the shock, large flights of seabirds flew inland over the city. When it struck, horses stood with their legs spread out, heads down, and trembled. Aftershocks were preceded "by a rumbling, sub-terranean noise, like distant thunder" or like "many pieces of artillery." The bay bubbled and there was "a most disagreeable sulfurous smell." The land was uplifted ten feet, as later determined from the beds of dead shellfish.

The naturalist looked about the razed port and city—an earthquake, fire, and tsunami having destroyed the inhabited areas—and concluded: "Earthquakes alone are sufficient to destroy the prosperity of any country." Darwin thought volcanic activity and earthquakes were related. The city had been destroyed before and would be rebuilt again.

Eleven years later the noted English geologist Sir Charles Lyell arrived at the rude settlement of New Madrid on the west bank of the Mississippi to have a quick look around. Lyell was to geology what Darwin was to biology. Both traveled widely to observe what they later wrote about. The geologist's seminal work, *Principles of Geology*, had been in print for sixteen years by the time he arrived in New Madrid.

When Sir Charles and his wife debarked from the steamboat in the spring of 1846, the single inn was closed because of lack of business. They lodged with a German baker—"a man of simple manners"—and his family. Thirty-four years after the New Madrid earthquakes, the evidence of massive movements in the surface of the earth were still quite visible.

A guide showed Lyell "the sink-hole where the negro was drowned" and a dry lake bed that had been drained by the shocks. The geologist borrowed a horse and rode into the "sunk country." He "fancied" that he felt a small quake. This was later confirmed by others who heard a sound "like distant thunder" at the same time. After two days, the Lyells departed the gloomy settlement on a steamboat. Life on board the gay Mississippi River vessel, wrote the noted geologist, was "like the fiction of a fairy-tale" in comparison to the reality of New Madrid.

Theories concerning the pace of geologic transformations have veered between gradual or instantaneous change, the former being known as *uniformitarianism*, or *gradualism*, and the latter as *catastrophism*. Lyell was the champion of gradualism, and nothing he saw at New Madrid changed his mind.

By midcentury California had its first earthquake sleuth. A doctor with one year of medical school, John B. Trask had an abiding interest in geology. For being an amateur, he was later derided by the professionals.

At personal peril, the first state geologist traveled throughout California and collected information on minerals and earthquakes. Gradually he constructed a history of earthquakes. Trask was the first to publish a chronological list of temblors and thus alert others to this important historical legacy. Given the complex geology and a region whose counterpart on the East Coast extended from Maine to Virginia, his biographer noted: "One cannot but be awed by the energy of this nearly forgotten nineteenth century physician-naturalist."

In September of 1869, the University of California opened its doors and higher education became formalized in the state. Michael L. Smith wrote in *Pacific Visions: California Scientists and the Environment, 1850–1915*: "Although scientists in California tried to keep abreast of theoretical disputes, their history reveals more about California, and about the social role of the scientist there, than about the history of scientific thought." In other words, the role of scientists was first to consider the needs of a growth-oriented culture.

Joseph Le Conte was the university's first professor of geology, botany, and natural history. Educated at Harvard University under Louis Agassiz, one of the lions of natural history, he arrived at the start of classes with the latest theories about earthquakes. These eastern theories, derived in turn from Europe, did not match the realities of life in the nation's most earthquake-prone state.

Hard on the heels of the numerous earthquakes and eruptions of Mount Vesuvius and field studies in Italy by Robert Mallet, an Irish engineer credited with making the first detailed, scientific investigation of seismic events, the prevailing theory linked earthquakes to volcanoes. Trask and Le Conte subscribed to that theory.

In a series of articles in the university newspaper in 1872, Le Conte said that up to twenty years ago there had been no scientific knowledge of earthquakes, just "rubbish in the form of loose popular observations." (So much for Darwin and Lyell.) One of the problems for the careful researcher, he said, was the suddenness of earthquakes "and the terror they inspire utterly unfitting the mind for scientific observation." To Le Conte, earthquakes were a vertical phenomenon.

That same year there was a great earthquake on the east side of the Sierra Nevada that left a gaping fault line nearly one hundred miles long. Horizontal displacements reached twenty-three feet. The shock stopped clocks and awakened people from San Diego to Elko, Nevada, and greatly alarmed John Muir in Yosemite Valley. Almost all the structures in Lone Pine were destroyed, and there were twenty-seven deaths—a devastating experience for a small community. Destruction and deaths also occurred elsewhere in the remote Owens Valley.

Josiah Whitney arrived to take a look two months later and found "a scene of ruin and disaster." The residents had heard sounds, "like artillery going off in rapid succession." The sounds were followed by violent shaking. Trees and boulders crashed down the steep slopes of the fourteen-thousand-foot peaks of the Sierra Nevada. There were the usual stories of cattle being "squeezed to death in the fissures." Whitney published his findings in a popular magazine. He said that there was no way such events could be predicted. The best protection, Whitney concluded, was "building in a suitable manner."

NEW YORK CITY AND CHARLESTON

The East Coast was not immune to seismic disturbances during the nineteenth century. There had been a slight tremor in New York City in 1737, but it had been forgotten by 1884. New York's greatest earthquake—albeit one of quite moderate size—occurred on August 10 of that year. The epicenter was fifteen miles offshore of what is now the John F. Kennedy International Airport. The first shock came at 2:07 P.M. and was followed by two others that were felt from Maine to Virginia to Ohio.

The city's largest newspaper, the *New York Herald*, threw all its considerable resources into coverage of the great event. The headlines the next day declared:

A LIVELY SHAKE
New York Visited by
a Genuine Earthquake

THE CITY'S NOVEL EXPERIENCE
Rumblings, Mutterings and Tremblings that
Greatly Alarmed the People

When the quake first hit, accompanied by a "hollow, rumbling sound," residents thought it was an explosion. Panic struck when they realized it was a temblor. The newspaper reported:

> In many of the thickly populated tenement house districts the shock caused great apprehension. All over the city people ran from their houses, fearing the collapse of the building. They threw household furniture from the windows or dragged their valuables down the stairs after them. A number of rumors of calamities sprang up within fifteen minutes of the shock.

Brick chimneys collapsed, fruit was shaken from trees, river water became choppy, glassware broke in restaurants, walls of houses cracked, and the police thought that anarchists were on the loose. The *Herald* reported:

> The Police Headquarters building trembled to its foundation and all the officials on duty, headed by ponderous Captain Saunders, ran into the street, thinking that the massive structure was tumbling. Recollections of the recent dynamite explosion at Scotland Yard at first passed through the minds of the frightened Central Office officials, who supposed the rear, or Mott street, wing of the marble palace, had been blown up by malcontents.

Former Californians living in New York were more blasé. Said one: "I am used to earthquakes, having resided on the Pacific coast a number of years.

The moment the shock declared itself I was sure it was an earthquake. To verify its duration I whipped out my watch and, though my little daughter was in hysterics, I timed the disturbance."

There was widespread terror but only light damage. One man died "from fright" in the Hartford, Connecticut, jail.

The *Herald*, like West Coast newspapers, reassured its readers:

> It needs no theorizing when such tremendous earthquake velocities are realized in a comparatively yielding medium, to see that the subterranean forces which disturb the serenity of mankind have never yet been overrated, and our confidence in *terra firma*, even on our own favored continent, is not wholly unshaken. But, judging from the history of our earth shocks, there is no reason to fear an immediate recurrence of Sunday's paroxysm on the Atlantic coast.

The New York Times comforted its readers in a similar manner. Newspaper coverage was intense for a couple of days, then slacked off quickly.

Two years later walls again cracked and chimneys fell in New York, this time the result of a strong earthquake in Charleston, South Carolina—the East Coast's greatest temblor in historic times.

The late summer afternoon of August 31, 1886, was unusually sultry. Only a slight breeze played across the charming southern city known for its churches, gardens, tiered piazzas, and erratic street patterns constructed over three centuries. The curator of the local museum, a physician and natural historian, noted:

> As the hour of 9:50 was reached, there was suddenly heard a rushing, roaring sound compared by some to a train of cars at no great distance, by others to a clatter produced by two or more omnibuses moving at a rapid rate over a paved street; by others again to an escape of steam from a boiler. It was followed immediately by a thumping and beating of the earth underneath the houses, which rocked and swayed to and fro. Furniture was violently moved and dashed to the floor, pictures were swung from the walls and, in some cases, completely turned with their backs to the front, and every moveable thing was thrown into extraordinary convulsions.

Hardly a structure in the colonial city was not damaged, and sixty persons were killed. Scattered fires broke out across the city but did not unite to form a single conflagration. Damage was greater on "made ground." Swaying wooden structures held up better than rigid brick buildings.

Charleston fit the San Francisco model in other ways. Robert P. Stockton wrote in his history of the quake: "Charlestonians were accustomed to disaster. They had a history of it. The city had survived hurricanes, tornadoes, bombardment and occupation by hostile armies in two wars, devastating fires, pestilence, economic depressions, and earthquakes in the past." The city was quickly rebuilt.

For the first time the USGS carefully investigated an earthquake. Four thousand reports were collected from sixteen hundred locations and were subsequently cataloged. Yet the cause remained a mystery. Clarence E. Dutton summarized the findings: "But after the most careful and prolonged study of the data at hand, nothing has been disclosed which seems to bring us any nearer to the precise nature of the forces which generated the disturbance."

BIRTH OF THE CONCEPT OF FAULTING

There was an important shift in seismic thinking and attitudes as the end of the nineteenth century approached. The intraplate quakes of New Madrid, New York, and Charleston were not near volcanoes. Scientists began to search for other explanations, and suspicion shifted from volcanoes to faulting along short blocks. Part of the reason for this change was that in the two decades before 1906 there was an increase in quakes on the San Andreas Fault. Accordingly, what became known as faulting was studied carefully.

Seismographs appeared on the scene at this time. The first seismographs installed in the Western Hemisphere—those at Lick Observatory, east of San Jose, and on the Berkeley campus of the University of California—were put under the control of astronomers in 1887. The explosion at a Berkeley gunpowder factory was the first seismographic record of ground movement in California.

From this point on, slowly at first and then with increasing speed in the next century, seismology based on direct observation in the field gave way to laboratory seismology that was dependent upon instruments and numbers. Quantification gave the practitioners of the new science credibility among their peers in the more established sciences.

Clarence Dutton was one of the giants of the early USGS and no slouch with words when it came to describing the Grand Canyon in his earlier years. He later wrote dismissively of penned accounts: "With most of the race, knowledge of [earthquakes] is derived wholly from written accounts, and these, until seismology became a science, were vague and more or less imaginative."

Dutton could have been referring to one such account published in the *San Francisco Bulletin* following a tremor on March 31, 1898. The anonymous author was relaxing on the top story of the new six-floor Press Club Building when the quake struck. Plaster fell from the ceilings, and walls cracked. The reporter, writing in a style reminiscent of Mark Twain, noted:

> Some rather humorous scenes occurred in the rooms during the quake. About twenty persons were lounging about the various rooms, and in the card rooms an interesting poker game was going in full blast. A newspaperman, who is famous for his Chinese plays, was sitting behind a full hand with a big jack pot in front of him. Opposite him was the natty attaché of a Federal court. The race track reporter of one of the big dailies was peering into aces up and a Judge from the City Hall was trying to figure out whether it would be a good idea to "come in" on a pair of deuces.
>
> Suddenly the building commenced to tremble. Then it shook. Sounds of crashing glass and tumbling bric-a-brac were heard. The man with the full hand jumped up in the air and exposed three kings and a pair of tens. The Federal man turned pale, but didn't lose his nerve. He carefully laid down his aces up, spread a pair of gloves over them, and then grabbed the arms of his chair as if he expected it would get away from him.
>
> The playwright started for the door and in the meanwhile there was great excitement in the library. Two large turned granite balls had fallen from a high shelf and nearly brained a student of Shakespeare. The student started for the same door that the man with the full hand was trying to get through and a collision occurred. Both pushed on the door desperately but as neither would give in they were forced to head via different routes for the elevator shaft. The elevator was at the bottom floor and the gentlemen took to the stairway and rushed into the street.
>
> The Federal man separated himself from his chair after the trembling, arose, smoothed the wrinkles out of a plug hat, went to the elevator shaft and shouted: "I call you; what've you got?"

The seismological community developed a distrust for such anecdotal accounts, even when less fanciful. Indeed, the very word *anecdotal* became a pejorative in the rush to numbers. In the process, earthquakes were dehumanized.

Modern seismologists like to claim their predecessors "discovered" the San Andreas Fault, an argument that resembles the claim that Columbus discovered America. What the Native Americans thought of that long rift valley is not known. Certainly the Visalia newspaper editor had a handle on the concept. Harold W. Fairbanks, a USGS geologist, traveled four hundred miles along the fault shortly before the 1906 quake. He wrote, "As a matter of fact, certain portions of the great rift have been known to the country people living along it for many years."

In the last decade of the nineteenth century, a number of geologists, most prominently Andrew C. Lawson of the University of California and John Casper Branner of Stanford University, nibbled away at the concept of one fault, detecting a piece here and there. In his history of the fault's discovery, Mason L. Hill wrote that the various accounts "show that these early field geologists, like the blind man guessing the nature of the elephant, were really in the dark with respect to the fault's length."

The nod for "discovery" is usually given to Lawson. It should actually be a slight dip of the head, given the short distance he identified the fault on the San Francisco Peninsula and the fleeting mention he gave it in two publications in the last decade of the nineteenth century.

Other geologists filled in some of the broader pieces of the seismic puzzle. Through his work in Utah and the Great Basin, Grove Karl Gilbert suggested that a vertical fault mechanism was the source of shocks. He wrote in an 1883 issue of the *Salt Lake Tribune*: "In yielding to this all-compelling upward thrust, the earth's crust sometimes bends and stretches, but more often it breaks; and when it breaks, the fracture occurs in a peculiar place."

Gilbert posed a crucial question and supplied the correct answer: "What are the citizens going to do about it? Probably nothing."

The term *dislocation quakes* was used by Dutton, who wrote a book on earthquakes published in 1904 that was considered the most comprehensive at the time. Charles Davison, an English geologist, wrote in 1905: "That tectonic earthquakes are closely connected with the formation of faults seems now established beyond doubt. They occur far from all traces of recent volcanic action."

Theoretical science was now prepared for the startling event that was about to unfold.

6 SAN FRANCISCO
(1906)

5:12 A.M.

Grove Karl Gilbert was asleep in the faculty club at the University of California on April 18, 1906, when the San Andreas Fault ruptured. He was awakened by the shaking at 5:12 A.M. The geologist wrote:

> It is the natural and legitimate ambition of a properly constituted geologist to see a glacier, witness an eruption and feel an earthquake. The glacier is always ready, awaiting his visit; the eruption has a course to run, and alacrity is needed to catch its more important phases; but the earthquake, unheralded and brief, may elude him through his entire lifetime. It had been my fortune to experience only a single weak tremor, and I had, moreover, been tantalized by narrowly missing the great Inyo earthquake of 1872 and the Alaska earthquake of 1899. When, therefore, I was awakened in Berkeley on the eighteenth of April last by a tumult of motions and noises, it was with unalloyed pleasure that I became aware that a vigorous earthquake was in progress.

The redwood-planked faculty club creaked; the furniture and bric-a-brac rattled. Gilbert noticed as he lay in bed that he was being tossed in a north-south direction. The suspended light fixture also swung in a north-south arc and water in a pitcher sloshed to the south. Enough confirmation for a firm conclusion? Not really. In another part of the faculty club, an east-west motion predominated. The lesson was, be careful; this was a complex and unpredictable phenomenon.

Less enchanted with the strange goings on was Clarence E. Judson, who lived in Ocean Beach on the San Francisco coast. Judson was taking his customary early-morning dip in the frigid ocean. He was the one person most directly exposed to the immediate force of the earthquake. The epicenter was just a couple of miles offshore of San Francisco.*

The surf was moderate that morning, but the breakers did not advance toward the beach in their usual pattern. Judson said, "They came in crosswise and in broken lines with a vicious, snappy sort of rip-and-tear fashion." The chop was caused by foreshocks.

Judson was up to his shoulders in the water when he was struck by a large breaker that ran up the beach a fair distance—not an unusual occurrence. But what followed was quite out of the ordinary:

> It almost took me off my feet, and I started to go out, and instantly there came such a shock. I was thrown to my knees. I got up and was down again. I was dazed and stunned, and being tossed about by the breakers, my ears full of salt water and about a gallon in my stomach. I was thrown down three times, and only by desperate fighting did I get out at all. It was a close call.
>
> The motion of the quake was like the waves of the ocean—about twenty feet between crests—but they came swift and choppy, with a kind of grinding noise—enough for anyone.
>
> I tried to run to where my shoes, hat and bathrobe lay, but I guess I must have described all kinds of figures in the sand. I thought I was paralyzed. Then I thought of lightning, as the beach was full of phosphorous. Every step I took left a brilliant incandescent streak. I jumped on my bathrobe to save me.

What Judson experienced was a type of luminosity that accompanies some seismic events.

---

* The epicenter for the 1906 quake was first thought to be at the place of maximum displacement at the end of Tomales Bay in Marin County but was later determined to be a point on the fault line just west of San Francisco. The epicenter is not necessarily the place of greatest movement but rather the point on the surface that is directly above the focus. The focus, or hypocenter, is the initial point of rupture within the earth.

He had difficulty getting dressed. Aftershocks kept throwing him to the ground. "I reeled and staggered like a drunken man. I thought of my wife and babies; I had left them asleep. I realized we had just had a terrible earthquake."

Judson ran home to find a badly frightened family. There was a tidal wave scare, and they camped in the sand dunes for a few days while the city burned to the east of them.

TO THE NORTH

The seismic waves sped across the water at speeds of up to three miles per second.

A number of ships were plying the coastal waters near the epicenter. A pilot boat, lying near the lightship that marked the San Francisco Bar, quivered. The captain of an inbound German steamship said it felt as though his ship had struck rocks, and others reported the same hard jolts. The tidal gauge at the Golden Gate registered a four-inch drop, but there was no tsunami.

To the north, the chief engineer of a steamship said, "The ship seemed to jump out of the water; the engines raced fearfully, as though the shaft or wheel had gone; then came a violent trembling fore and aft and sideways, like running at full speed against a wall of ice. The expression 'a wall of ice' is derived from my experiences in the Arctic."

On land along the San Andreas Fault—first to the north, then to the south, and lastly to the east in this account—destruction and death were almost instantaneous in the early-morning hours of that spring day.

Northward from off the Golden Gate, the fault comes ashore at Bolinas. To the west the land was slightly uplifted, to the east it dropped. Buildings on stilts along the shore of Bolinas Lagoon slid into the water. At the Flag Staff Inn "one may sit along the upper edge of the parlor floor and fish in 4 feet of water along the opposite edge of the same room," reported a professor. The channel inside the lagoon shifted, and as a result some vessels went aground. People were thrown from their beds, but no one was seriously injured. Structures on higher, more solid ground fared better.

Further toward the northwest, at Point Reyes Station, a steam engine and three railroad cars were overturned. The school was damaged, and a brick

store collapsed, as did most chimneys in the railroad and ranching community. Between Point Reyes and Inverness, Gilbert measured a twenty-foot
offset on the levee road, now known as Sir Francis Drake Boulevard.

Illustrating in a dramatic manner that great movement does not necessarily mean great proximate damage—that there are other important
factors—the ground shifted fifteen feet, nine inches immediately in front of
the Skinner ranch, now the headquarters of Point Reyes National Seashore.
The wooden structure was barely damaged; the path, however, no longer led
to the front steps.

In the nearby village of Inverness half the houses shifted on their foundations, water tanks slipped from their perches, and piers that reached out into
Tomales Bay were bent as much as thirty feet. At the lighthouse on the
point the heavy mechanism slid off its base. One of the keepers reported
the water "boiled" in the ocean. Along the coast landslides blocked roads.

Gilbert explored the fresh breaks along Marin County's portion of the
San Andreas Fault during the following weeks and added revealing details.
The clarity of Gilbert's observations and photographs, published two years
later in the state report, was in the tradition of the explorers and geologists who probed the American West and reported on their findings in
handsomely printed volumes during the nineteenth century.

The geologist was uncertain about the accuracy of his twenty-foot measurement of the road offset, which has now become embedded in earthquake lore as the greatest offset caused by that temblor. Gilbert's concerns
were: He wasn't sure if the road had originally been straight, he had measured a single wheel track that had been disrupted by fifty or sixty feet of
churned and sunken roadway, and he thought part of the distance could be
attributed to slippage of the firmer roadway over the squishy underpinnings
of the marsh. Gilbert was bothered by the fact that the twenty feet was
greater than the fifteen- to sixteen-foot displacements of the path, a fence, a
barn, and a row of raspberry bushes on drier ground one mile to the southeast at the Skinner ranch.

Gilbert traced the route of the fault line northwest across the road and
into the marsh at the foot of Tomales Bay, which subsequently became the
flat pastureland that is visible from my deck. His photos show an irregular, narrow channel traversing the pickleweed-dominated tidelands. He
described the cracks in the marsh thusly: "In general they were not parallel

with one another nor were they otherwise systematically arranged, except that some of them were apt to occur along the boundary between alluvium and a firmer formation."

The geologist explored the marsh on foot: "A large portion of the delta was thrown by the earthquake into gentle undulations, the difference in height between the swells and hollows being usually less than a foot," he wrote. Gilbert thought the corrugated surface was sculpted by the passage of the seismic waves. He noted a number of landslides around the bay and the fact that its muddy bottom had shifted westward.

Gilbert, the most noted American geologist of his time, was not infallible. Still recalled by geologists as a warning against giving credence to anecdotal evidence is the buried-cow episode at the Shafter ranch in Olema, just south of Point Reyes Station.

The story falls squarely within the category of earthquake apocrypha that dates back to biblical times and the apparitions of the earth opening and swallowing cities. Gilbert fell for the story, as did David Starr Jordan, the president of Stanford University and the author of a popular book on the earthquake, and a number of newspaper reporters.

The geologist reported that a cow was swallowed by a crack, leaving nothing visible but a tail that was eaten by dogs. He didn't actually see the cow or the tail but stated that "the testimony on this point is beyond question." Gilbert searched for a crack that was large enough to accommodate a cow but could find none. He attributed the incident to "a temporary parting of the walls."

While some locals said they saw the buried cow and others said they heard about it, the episode was most likely a practical joke played by Payne Shafter on the media. Shafter had a dead cow to bury, and along came the earthquake. He buried the cow and made up a good story for the bothersome newspaper reporters and geologists.

More serious events occurred in the town of Tomales to the north. Two girls were killed in a stone house. The Catholic church and a saloon, along with other structures, were completely demolished. Almost all the monuments in the cemetery fell in either a north or south direction.

From the mouth of Tomales Bay the new fracture sliced through Bodega Bay and was a visible fissure when it emerged from the ocean near Fort Ross. Above the settlement, originally built by the Russians, a six-foot-diameter redwood tree was cleaved in half and offset by eight inches. The offsets along the fault line reached fifteen feet in sparsely settled coastal Sonoma County, but the greatest damage was twenty-five miles inland at Santa Rosa.

Proportionally, everyone agreed, Santa Rosa, with 6,700 inhabitants, suffered the greatest damage of any city. Author Jack London, who visited Santa Rosa and San Francisco at the height of their respective conflagrations, said, "Santa Rosa got it worse than S.F." A 1907 USGS assessment of quake and fire damage came to the same conclusion. In both cities fire followed the earthquake like a malevolent twin. And in both cities history was obliterated or rewritten to an extent matched only later in the century by various totalitarian regimes.

An extensive history of Sonoma County published thirty years later contains but two fleeting references to the earthquake and fire. In a pamphlet filled with searing photographs of destruction rushed into print in the heat of the moment and before the cover-up, J. Edgar Ross wrote, "The tremor lasted less than a minute, but in that time it converted the beautiful City of Roses into a City of Ruins." Of one building where four died, Ross commented bitterly: "Four lives sacrificed to the great American passion—a desire to get the largest possible returns from the smallest possible investment." There were more than sixty deaths in Santa Rosa.

Indeed, it seemed that either ignorance or avarice was to blame. True, Santa Rosa was built on alluvial soil deposited by creeks flowing from the Sonoma Mountains, about as poor a foundation as the made land surrounding San Francisco Bay. But shoddy design, workmanship, and materials were the principal reasons for such massive destruction at so great a distance from the fault.

Charles Derleth Jr., a professor of structural engineering at the University of California, visited Santa Rosa after the quake. He noted, "Its business district was wiped out by fire and practically every brick building in the business district collapsed in the earthquake." How could that be, so far from the fault? Derleth answered his own question: "The brick buildings of Santa Rosa were carelessly constructed." The mortar between the bricks was the consistency of sand. He added that the city burned, not for lack of water, but because of inadequate fire-fighting equipment.

The local newspaper exhorted its readers to have "faith in the future" and rebuild quickly. A few could not forget the past. Gilbert visited the noted horticulturist Luther Burbank at his farm in nearby Sebastopol, which had also been hard hit. Burbank recalled that during the thirty-plus years he had lived in the Santa Rosa area he had felt 130 earthquakes.

The newly etched fault line paralleled the Sonoma coastline a few miles inland until it reached Alder Creek, just north of Point Arena. The 110-foot brick lighthouse on the point was cracked in five places and subsequently condemned. The keeper on duty at the time said: "A heavy blow first struck the tower from the south. The blow came quick and heavy, accompanied by a heavy report. The tower quivered for a few seconds, went far over to the north, came back, and then swung north again, repeating this several times."

A bridge was destroyed at the mouth of the creek, and the coastal bluff overlooking the ocean was "marked by characteristic rending and heaving of the sod." From the beach the fault dropped into the ocean to appear for the last time on land at Shelter Cove, where it pierced the tip of Point Delgada before diving into the complexities of the Mendocino Triple Junction.

FROM SHELTER COVE TO FORT ROSS

The portion of the San Andreas Fault I chose to follow in order to present a contemporary perspective on the 1906 temblor extends from Shelter Cove in Humboldt County to Bolinas in Marin County. My journey from north to south was the reverse of the direction of the shock waves.

The northern terminus of the San Andreas Fault begins in fog- and storm-shrouded obscurity just off the Lost Coast, a roadless portion of the heavily forested mountainous region near the Oregon border. Eight hundred ten miles to the south, under a blazing sun and just beyond the flat, treeless Colorado River delta, the strands of the fault system sink into the semi-tropical waters of the Gulf of California. The 1906 quake tore the land apart for 260 miles from Shelter Cove, the one inhabited spot on the Lost Coast, to the mission town of San Juan Bautista in central California.

I have been at the system's extremities a number of times and have never ceased to wonder at their surface disparity. There is, however, a similarity; for I have felt the cobble on the beach at Cape Mendocino and the fine silt of the delta shake briefly underneath me. Those were minor quakes, common to both areas and the land that lies between them.

On the wall of my office, just to the left if I look up from the computer screen, is a vividly colored USGS/University of Nevada, Reno, map entitled "Earthquakes of California and Nevada." The same map is on the walls of the earth science buildings at the University of California at Berkeley and Caltech and various USGS offices. Epicenters are marked by red dots. The thickest swatch of red in northern California lies just off the coast between Punta Gorda and Cape Mendocino.

This is where the San Andreas Fault merges with the Mendocino Triple Junction. There is almost constant unrest where three tectonic plates—the Gorda, Pacific, and North American—meet and where there is a major change in the direction of crustal movement. The plates no longer slide horizontally past each other north of Punta Gorda but rather the Gorda and San Juan de Fuca Plates sink beneath the North American Plate.

Along the Northwest Coast there have been huge earthquakes in the past. The last occurred about three hundred years ago. Such giant quakes encompassing the entire Cascadia Subduction Zone, which extends from Cape Mendocino to Vancouver Island in British Columbia, were equal in magnitude to the world's greatest recorded shocks. The evidence for one massive shake, or a series of smaller temblors between 1700 and 1730 that achieved the same total effect, comes from such divergent sources as the radiocarbon dating of earthquake-stunted vegetation, trenching, and Yurok Indian tales.

At its northern end the San Andreas Fault touches land for the first time at Shelter Cove. The encounter is brief. Talk to people there, and they will tell you one of two stories, depending on their involvement in real estate sales:

- We have no earthquakes here, just a little shake now and then that doesn't really bother anyone.
- It's always shaking here, but we haven't felt anything in six months. I am really getting nervous.

The recent history of Shelter Cove has revolved around buying and selling the surface of the earth with little concern for what goes on underneath it. At this remote end of the continent—after a fleeting history of logging, ranching, and fishing had come and gone—a group of Los Angeles investors in the mid-1960s purchased five thousand hilly acres and divided the flatter pieces of ground into forty-six hundred tiny lots. They brought in tons of asphalt and paved forty-three miles of roads and a long runway atop an oceanside cliff. A nine-hole golf course was wrapped around the airstrip. The object was to sell land, not build a viable community.

Sales faltered. Only a few homes were built—thirty by 1973, when I first visited Shelter Cove. The roads, where torrential winter storms with gale-force winds dump more than one hundred inches of rain, were already starting to deteriorate. When I returned twenty-three years later, a real estate agent's pamphlet boasted: "Currently there are approximately 330 homes in the cove and it is increasing every day, just ask the contractors."

There were many realities not being addressed at Shelter Cove. Chief among them was the presence of the San Andreas Fault. The real estate agent's brochure certainly did not mention its presence, no sign pointed to its existence, nor did the one person who has lived there longest attach much importance to it.

Mario Machi moved to Shelter Cove from San Francisco in 1946. Machi, in his eighties, owned the prime commercial site that surrounded the boat-launching ramp. A former elementary school teacher, he was of the "doesn't bother me a bit" group. Machi wrote a book about the history of Shelter Cove that eventually got around to a brief mention of seismic activity:

> Although slight quakes have occurred in the past seventy years, there has been much controversy over the effect of earthquakes at Shelter Cove. The San Andreas Fault runs up the coast and passes through the subdivision. Earthquakes are unpredictable and everyone is aware of the forces involved. They should be placed in the same category as any accident that can occur in a home or automobile. There is no doubt that the danger of an earthquake is always present.

I drove past numerous For Sale signs planted in the front yards of homes and on empty weed-choked lots. The road ended at Telegraph Creek and Black Sand Beach at the north end of the subdivision.

A branch fracture from the 1906 quake slices through the mouth of Telegraph Creek. The main fissure lies upstream from the beach at the point where the county dump and water-treatment plant are located. Telegraph Creek is the source of water for the subdivision. The main fault zone passes over Telegraph Peak and emerges on the black sand beach just north of the mouth of Horse Mountain Creek.

I hiked to this point from the parking lot and stood atop a large boulder near the mouth of the creek. From there I had an unobstructed view in all directions. It reminded me of other places and times.

The peaks of the King Range were cut off at the knees by the gray coastal fog, much as the mist clamps down on the ridge that lies across the fault from my home. The ganglia of tree roots on the edge of the bluff that had been exposed by winter storms reminded me of the dead trunks of trees lying on the shoreline of Lituya Bay in southeast Alaska, uprooted by a massive wave caused by an earthquake. Here were the same black sand beaches found on the earthquake-prone Big Island of Hawaii.

I returned to my car and followed the fault south past deserted culs-de-sac where weeds have forced their way through the asphalt. I circled around the dense, fern-choked creek bottoms of the North Coast rain forest and rejoined the fault at lot 47 ("Make Offer"). Down the cleft that intersects Dead Mans Gulch at the south end of Shelter Cove runs the fault line.

Why Dead Mans Gulch?

Perhaps because on August 17, 1931, Charles East was riding his horse along the black sand beach in the cove looking for stray sheep in a heavy fog. The mist parted momentarily, and he saw a commercial fishing boat lying on its side in the shallows at the mouth of the gulch. There was a body in the wheelhouse. A local fisherman, whose name survives only as Anderson, had apparently died of a heart attack. Some years earlier this same Anderson had delivered the body of a fish and game warden, who had attempted to board his vessel, to the authorities in San Francisco.

Like attempting to decipher the movements of the earth, many details of this story are missing and will never be known.

I walked down the same beach on a similar foggy evening. Six surfers were riding the waves off the mouth of the gulch. Two others stood by their pick-ups, parked on the black sand. Dogs ran about. Other surfers were washing in the fresh water that flowed down the creek in Dead Mans Gulch.

I asked them about surfing on the fault. They said this was one of the best spots around. What made it so good, they said, was the reef formed by the rocks and sand flushed down the fault line by winter storms and deposited at the mouth of the gulch. The waves curled perfectly over the shallows in the summer twilight.

From Dead Mans Gulch to the mouth of Alder Creek just north of Point Arena is a distance of seventy-five miles. The fault zone lies just offshore and follows the curvature of the spectacular rock-studded shoreline past some of the most expensive coastal real estate in North America and such thriving communities as Fort Bragg and Mendocino, both heavily damaged in the 1906 quake. I think of this stretch of the fault, with its many precious tourist facilities, as bed-and-breakfastville.

The only indication at Alder Creek that the San Andreas Fault came ashore at this point was the temporary "Earthquake Retrofit" construction sign on the bridge over Highway 1. An earlier bridge over the creek had been destroyed by the 1906 quake. At the restored lighthouse at nearby Point Arena the nonprofit organization that conducts tours has a fault-line map and exhibits showing damage to the beacon and other structures in the area.

The geologists refer to the forty-five miles from Point Arena to Fort Ross as the Gualala Block. The various chambers of commerce, looking at it from a different perspective, call it Mendonoma, a bastardization of the names of the two counties it encompasses—Mendocino and Sonoma.

From the air over the Gualala Block the two plates can be easily recognized as separate entities; on the ground the geology is more difficult to comprehend. This portion of the San Andreas Fault has been studied the least. Parts of it are extremely difficult to reach. Roads cross the fault line at infrequent intervals one ridgeline over from the coast, and the thick vegetation is impenetrable in places. The Garcia River and the Little North Fork, the North Fork, and the South Fork of the Gualala River define the fault zone through most of the Gualala Block.

The local newspaper said that "Mendo" and "Noma" met at the Gualala River, so I set off to find this mythical coupling. The meeting place was on the fault line at the junction of the North and South Forks of the Gualala River.

The first coastal ridge blocked the cool summer fog and the gusting northwest wind. It was warm and sunny where I stood on the temporary wooden bridge recently constructed at the mouth of the North Fork by a logging company. Nearby was a green steel-girdered bridge that had once served a railroad and now accommodated one-way vehicle traffic.

Between the two bridges pink surveyor tape drew attention to the two pages of the Green Bridge Flat Timber Harvest Plan nailed to a tree trunk. Gualala Redwood Incorporated was logging one hundred acres adjoining the South Fork.

While I took in the peaceful scene of a summer river snaking past exposed gravel bars on the fault line, a pickup truck drove by hauling a length of culvert and raising dust in its wake. Then a white sports utility vehicle stopped. A woman, four children, and two dogs piled out and inspected me and the river.

"Is this where the fish are biting?" the woman asked.

I replied that I was no fisherman, rather a researcher of earthquakes.

"No kidding," she said. The woman left the children and departed momentarily to park her vehicle.

A second pickup stopped. A man leaned out of the window: "If everything goes all right, there should be some logging trucks coming by in a while." He drove off in a cloud of dust.

The oldest girl, who was standing beside me, commented dryly that the passage of the trucks might resemble an earthquake.

I told the woman about the logging trucks when she returned, and we chatted while the two boys fished and the dogs and the other children raced in and out of the water.

Judy Findeisen of San Jose was an exuberant fifth-generation Californian who was turned on by earthquakes. "I love them," she said. "We love them: the unknown about it; how big it's going to get; when it's going to stop; how long it's going to last; what's going to fall down. They are unpredictable. And they get your blood pumping."

While standing on the gravel bank and talking, I had noticed a truck pass with a colorful assortment of canoes, kayaks, and life preservers attached to racks. Two kayakers waded up the river, pushing their craft over the shallows. From my map, I knew a bridge crossed the north-flowing South Fork some seven miles distant.

Could I run the San Andreas Fault in a canoe or kayak? The water was obviously too low in July, so I returned the next spring. I had waited too long. Although it was only the beginning of April, the river was already quite low. Huge rains in January had caused massive flooding, but very little rain since then had depleted the flows. The same gravel bars were again visible.

My friend Michael Mery accompanied me in a canoe. Wayne Harris, who operated a local canoe and kayak rental business, said very few of his customers were aware that the South Fork of the Gualala River was the San Andreas Fault.

We put in at the Annapolis Road bridge for the nine-mile run to the Highway 1 bridge. It should take us three to four hours, Harris said. Two kayakers who had done the same trip a day or two before had to get out and walk their craft over the shallows a half dozen times. "And by the way, watch out for the first turn. We had some canoers dump there last weekend, and they said it was a long, wet, cold trip downriver."

I said not to worry and cited my extensive experience on rivers, bays, and ocean waters of the West. My friend was a native of this coast and familiar with all types of small craft. This river with nary a rapid was a piece of cake.

We overturned in the first narrow bend of the river. I looked back, but Harris had departed. I was embarrassed. Fortunately, the day was warm, there wasn't much headwind, and we managed to stay right side up for the remainder of the trip.

When Harris picked us up four hours later, I confessed. He said he and a friend had dumped there, too. That made me feel a little better.

What wasn't much fun was having to get out and walk the canoe over the shallows some twenty or thirty times. Harris allowed that the river-running season was just about over for the year.

For those who want to take the trip in higher water, it is well worth it. Besides the fault line, we saw numerous ducks, a few deer, and one river otter. We passed a fisherman who was battling a steelhead trout. When not plodding through the cold water, we floated down the gentle channel that followed the relatively straight passage of the fault zone through the redwoods.

The most recent fissure follows the west bank and is quite obvious in places. In fact, the most visible portion is easily accessible by vehicle and foot.

The Sea Ranch subdivision has a picnic grounds along the river at a place known locally as the Hot Spot, conceivably because it is a lot warmer one ridge over from the ocean during the summer months, when the wind and fog buffet the tastefully designed redwood and cedar homes on the coast. At the end of River Beach Road the graded dirt road makes a loop around the recreation area. Just before it ascends, there is a trail leading north along the South Fork.

The 1972 USGS map of the fault line that I was carrying stated: "parallel trenches; largest on southwest." I followed a clearly defined trench for about a quarter mile, then picked up the parallel trench just beyond the point where the trail branched. The redwood forest, with a thick understory of ferns, had been logged ninety years ago; but the second growth, like a California city a few years after an earthquake, had recovered very nicely.

Sea Ranch, as did the more plebeian Shelter Cove subdivision to the north, drew its water from the fault line. The exclusive subdivision claimed to be "one of the most beautiful places in the world," but there was no mention of the San Andreas Fault in any of the promotional literature. However, in a history of the area distributed by Sea Ranch Realty I found the following explanation for the past logging activity that I had seen along the fault-carved river: "Following the 1906 earthquake and fire, San Francisco again looked to the North Coast for the lumber to rebuild."

Fort Ross, near the southern end of the Gualala Block, is a state park that preserves vestiges of the Russian presence in California but completely neglects the sag ponds, scarps, trenches, split redwood trees, and offset roads, creeks, and fences that mark the passing of the San Andreas Fault through public property.

The fault zone bisects the abandoned Russian orchard that lies behind a high wire fence just off Fort Ross Road, a half mile east of the stockade and at a point where a sign states "Stanley S. Spyra Memorial Grove." I don't know who Mr. Spyra was, other than a benefactor of redwood trees, but I do think that the fault deserves at least equal mention.

Although I'm not a gardener, the orchard has always fascinated me. It has survived for a long time at the exact point where the two tectonic plates grind against each other. Beginning in 1814 with seeds obtained from the Spanish in San Francisco, the Russians secured their starts from other earthquake-prone places: Monterey, California; Lima, Peru; and Hawaii.

When the Russians departed in 1841 there were 207 apple, 29 peach, 10 pear, 10 quince, and 8 cherry trees. In subsequent years the lack of blight on these trees, compared to similar varieties planted elsewhere in California, was attributed to the holy water sprinkled on them by a Russian priest.

Two miles south of Fort Ross, where Highway 1 jogs inland to Timber Gulch, the fault zone dips into the ocean, leaving behind a landslide-prone slope that has been repaired many times since Chinese laborers hacked a road out of the unstable face of the cliff in the 1870s. From the north, the left turn and gradual ascent across the most recent and most ambitious road repair job was once known as Earthquake Point or Black Bart's Turn.

On August 3, 1877, one year after the road was completed, a man dressed in a long linen duster with a flour sack over his head and a double-barreled shotgun in his hands stopped the Point Arena–Duncans Mills stage and commanded in a deep voice: "Throw down the box."

The driver complied. In return for the three hundred dollars in the box, the robber left a poem signed "Black Bart the PO8 [po-eight]."

The poem read:

> *I've labored long and hard for bread,*
> *For honor and for riches,*
> *But on my corns too long you've tred,*
> *You fine-haired sons of bitches.*

Four years later at the same spot California's least successful bandit, Dick Fellows, held up the stage. He opened the box and all it contained was a letter written in Chinese. Fellows was a klutz. He loved the grandeur of robbing stages on horseback, but he was a poor rider. Horses threw him or ran away with him.

Black Bart turned out to be a Civil War veteran by the name of Charles E. Bolton. He was captured and jailed in San Andreas, a county seat in the Sierra Nevada foothills.

The unmarked turn is worth a pause. Looking back from the opposite side of Timber Gulch at the massive slide is the closest I have come to being able to visualize the passage of the plates. The shelf containing the Gualala Block appears to be passing behind a slight ridge that serves as the point of slip for the two massive plates. The ridge drops through pulverized gray rock to the white surf that swirls about the black rocks at the foot of the steep cliff. The summer tourist traffic negotiated the curves on picturesque Highway 1, ignorant of the history of crime and tectonic movement.

THE POND

Toward the south I could just make out the low mound that marks Bodega Head. The fault comes ashore at Salmon Creek and then passes along the foot of the bluff in Bodega Bay, where the tourist facilities and commercial fishing docks are located.

The crooked arm of Bodega Head, a granite outcrop, forms the protected harbor where the Russians anchored their ships. Campbell Cove, within the crescent, is a peaceful wildlife preserve. Reeds grow in profusion and red-winged blackbirds nest around a pond fed by underground springs that was once the foundation for a nuclear power plant that was never built. It was here that a giant utility company was undone by an earthquake fault.

It helps to have a slight personal involvement in order to fully appreciate history. I was a young reporter on the daily newspaper in Marin County in the early sixties. The city editor assigned me to a Pacific Gas & Electric Company (PG&E) press tour of the nuclear power plant site. It was the first time that I rubbed elbows with reporters from the glamorous San Francisco dailies and was courted by public relations people from a large corporation. Needless to say, I was awed.

We traveled to the site by bus. A sign announced that we were entering "Bodega Bay Atomic Park," a sop, I now understand, to plans to make Bodega Head a state park in return for the industrial site. Workers wore aluminum hard hats that looked like World War I helmets. Huge earthmoving equipment crawled over the bare ground of Campbell Cove. A gaping hole 140 feet wide and 70 feet deep was corseted with thick steel ribbing.

The overconfident utility company had already spent millions of dollars and had begun to excavate the foundation for the reactor before final approval by the Atomic Energy Commission (AEC). There seemed no way to stop the powerful utility. The opponents were characterized as "little old ladies in tennis shoes," "Chicken Littles," and "Communists" in the derogatory vernacular of the time.

We were smothered with press releases, geologic facts, and engineering plans and taken to a liquid lunch, compliments of PG&E, the nation's largest private utility. We stumbled back to our respective offices to write stories favorable to our host.

The opposition, led by a young Berkeley activist named David Pesonen, released balloons from the site that floated downwind to the pasture lands of West Marin where, theoretically, grazing cows would produce contaminated milk. Pesonen had split off from the Sierra Club when such tactics, and his charges of collusion between public officials and utility company executives, angered the moderate leadership of the conservation organization.

The times, however, were changing. Civil rights, student unrest, Rachel Carson's *Silent Spring*, fallout from nuclear bomb tests, and the movement for a moratorium on further tests were transforming the modus operandi of citizen groups from polite negotiations to angry confrontations.

The most powerful institutions promoted nuclear power plants. Congress and the AEC were pushing the peaceful use of the atom. PG&E was facing rampant population growth within its service area and needed to expand. The utility already had two conventional power plants and one nuclear facility on the coast. Such sites offered plentiful and cheap cooling water.

The Bodega Head debate centered on seismic safety and was a classic encounter that set the pattern for others that followed. The utility hired the best consultants available from the most learned institutions. They included Don Tocher, who held a top position at the Berkeley seismological station; Hugo Benioff, the author of the earthquake entry in the *Encyclopaedia Britannica*; and George Housner, known as "the father of earthquake engineering." The latter two were from Caltech. All three said there was no seismic problem.

Pesonen's expert, Pierre Saint-Armand, a seismologist at the China Lake Naval Weapons Center, peered into the hole at Bodega Head and declared: "A worse foundation situation would be difficult to envision."

Forced into a highly visible public forum focused on a specific issue for the first time in the short history of their discipline, seismologists—like the fault itself—were badly fractured. As a historian for the Nuclear Regulatory Commission noted nearly thirty years later, "Geology, at least so far as understanding earthquakes was concerned, provided only contestable hypotheses rather than immutable truths."

Politics entered the fray and further bent the issues. Republicans generally favored construction of the plant; the Kennedy administration was against it. The issue was a minor addendum to the cold war. J. Edgar Hoover's FBI secretly tracked Pesonen's group, and PG&E's more extreme supporters and employees maintained the organization was a Communist front.

The USGS was asked to take a look. The site was given preliminary clearance by two USGS geologists who considered only the consequences of direct faulting, not the possibility of shaking. The area was prone to both types of movement.

The survey's senior seismologist was dispatched to the site. The opening words of Jerry P. Eaton's report were a devastating comment on the limitations of modern seismology and its inability to deal with realities on the ground.

He wrote:

> The primary difficulty is that the seismologist is called upon to make judgments that require large extrapolations beyond his personal professional experiences and even beyond those of the science he serves. When such seismological judgments are short of qualifications and condensed to a convenient statement for engineering guidance, they take on an unwarranted ring of certainty that belies their shaky foundations.

Eaton concluded:

> The case against the site stresses seismology's lack of detailed information on events and conditions in the epicentral tract of a major earthquake. Because we cannot prove that the worst situation will not prevail at the site, we must recognize that it might.

At the time, such caution concerning the siting of nuclear reactors seemed revolutionary. Today, Eaton's prophetic words are applicable to all manner of structures. Sadly, we no longer trust the experts. A historian recently commented on the Bodega Head controversy: "By revealing differences among experts, the uproar contributed to the loss of faith in technical elites and government generally."

Eventually a fault was found that bisected the reactor site, PG&E gave up its plans, Bodega Head became a state park, and not long ago my wife and I picnicked by the pond. Given the peacefulness, she found it difficult to imagine all that had taken place at this unmarked historic site.

Along the California coast seismic concerns subsequently delayed construction of other nuclear power plants at Diablo Canyon and San Onofre, and ruled them out at Point Arena, Davenport, and Malibu. In his pro-utility company book, *The Atom and the Fault*, Richard L. Meehan noted that "geological issues had nearly broken the back of the nation's nuclear industry."

There is an addendum to this story. A few years ago an older and wiser writer applied to the public relations department at PG&E for a tour of its nonfunctioning Humboldt Bay nuclear plant. I was told they weren't conducting tours. It wasn't difficult to guess why; there was little to boast about.

The plant, just south of Eureka on the Mendocino Triple Junction, went into operation in 1964, the year Bodega Head was abandoned. It was the nation's first commercially viable nuclear power plant. Twelve years later it shut down because of seismic concerns.

A moderate quake jolted the area in 1980. A nearby freeway overpass collapsed, and the radioactive water within the container holding the spent fuel rods came within inches of spilling. There were minor cracks in the concrete and masonry of the building. Fissures disturbed the land around the plant, which had become a symbol of the demise of the once great promise of nuclear power.

FROM TOMALES BAY TO BOLINAS LAGOON

I live twenty miles downwind from what would have been a nuclear power plant. From the mouth of Tomales Bay to Bolinas at the south end of the Olema Valley, the ancient rift is a dramatic dent in the surface of the earth.

There is a physical resemblance to Loch Ness, another seismically disturbed region. There have also been repeated sightings of a sea monster here by such persons as a sculptor, an architect, a lighthouse keeper, a commercial fisherman, and five state highway department employees. I frequently swim and row in the fault-formed bay—termed "Earthquake Bay" by a local historian—but I have never seen anything out of the ordinary.

My wife and I are twenty-year residents of the area. We have lived on both plates. To drive from one to the other is to change worlds—from the darker, cooler, more densely vegetated Pacific Plate to the light-filled, warmer, more open North American Plate. The transition occurs just about where Gilbert measured the twenty-foot offset one mile from our home.

We take chances living in such a place; and, like most Californians, we hope to slide by the many natural catastrophes that beset this state. The seismic odds seem to be in our favor. Our segment of the fault has been given a 2 percent probability rating for a major earthquake within the next twenty years. For the whole Bay Area, the probability is 67 percent. However, I don't put much credence in such numbers. They have a false ring of certainty about them.

In 1990 we purchased a one-acre lot on the North American Plate. The upper one-third of the narrow lot is flat. It then drops nearly one hundred feet to an abandoned railroad bed and Tomasini Creek. Beyond the creek is the fault zone in the form of a flat field where cows graze year-round.

The fault bisects the field that was once a marsh at the southern end of Tomales Bay before a rancher constructed levees and drained the water in 1946. When the rains come in the winter, the field again becomes a marsh. From fall to spring tens of thousands of migratory birds—teal, northern shovelers, and hooded mergansers, among others—feed along the fault line.

Blue herons and snowy egrets are more permanent residents. Deer, frequently does and their accompanying fawns, graze in the field, and once we saw a coyote jogging along parallel to the fault. The National Park Service plans to buy the ranch and restore the marsh. A more bucolic scene is hard to imagine.

Rather than build an intrusive structure on the flat area, we elected to wedge the house into the slope. We observed every safeguard mandated by law in this relatively earthquake-conscious state. Two soil surveys, one for the subdivision as a whole and the other for our specific lot, were made by the same geotechnical engineer. He dug a total of four test pits on our lot. The soil engineer saw no signs of recent landslides on our property. He noted that the site was underlain by "modest strength terrace deposits" and not bedrock materials.

The clay, silt, sand, and gravel came from the periodic advances and retreats of salt and fresh water over tens of thousands of years. An elephant's tooth and a ground sloth's humerus were discovered nearby in this same geologic formation, as well as signs of prehistoric earthquakes.

Noting that the proposed house was 1,400 feet from the San Andreas Fault zone, the engineer pointed out that should there be an earthquake, the alluvial soils would "increase or amplify the influence of bedrock ground vibrations." He added: "We recommend that the design engineer provide structural and foundation systems that emphasize the principles of ductility, continuity, and high energy absorption."

The structural engineer took this information and designed a drilled pier and grade beam foundation system consisting of forty-two reinforced concrete piers sunk twenty feet into the ground. The piers were tied together and the one-story wood-frame house was firmly attached to the foundation, whose cost amounted to one-third the total price of building the modest structure.

The thinking was that the foundation and superstructure would sway as a unit in any given earthquake. The Marin County building inspector signed off on the project, and we moved into our new home in 1991. What we did now, in regard to earthquake protection, was our responsibility.

First, we obtained an adequate amount of earthquake insurance, which was neither a requirement for the bank loan nor something most home-owners who live near the fault possessed. Earthquake insurance is not a given for everyone; there are a number of variables that need to be considered.

For instance, on our $300,000 earthquake policy there was a 15 percent deductible, meaning that we had to pay the first $45,000 worth of earthquake damage. Our home was new and had been engineered in excess of building-code requirements. Unless there was a huge earthquake centered nearby, which was unlikely according to the experts, our damages would probably fall within the amount of the deductible.

In 1996, we paid $513 a year for this earthquake rider from a private insurance company—in addition to the usual homeowners package that included fire, theft, and liability. Should the earthquake cause a fire, the damage would be covered under the more liberal homeowners policy, which has a $250 deductible.

The next year the state took over earthquake insurance because of the reluctance of private companies to offer such coverage after the huge losses incurred from the Northridge earthquake. The president of one California insurance company said in 1994: "Stated simply, California has a disaster problem."

The state rates were based on the amount of estimated seismic risk in postal zip code areas and the age of the structure. Since the risk where we live is supposedly minimal and we have a new home, our premium for the same amount of earthquake coverage was $300. The deductible remained the same. We were allowed $5,000 for personal property and $1,500 for living expenses, token amounts not included in our previous earthquake policy. Demolition and structural engineering costs were not covered, as they had been.

On the surface, it looked like a good deal. But in order to qualify for the same amount of earthquake coverage, we had to increase the amount on our homeowners policy, which also underwent some basic changes the same year. (That there were simultaneous modifications, to the benefit of the insurance industry, did not seem like happenstance to me.) The net result was we wound up paying a total of $135 more for slightly less coverage on both policies.

Additionally, we had to pay the earthquake premium in one lump sum, rather than in quarterly payments. When the "Basic Earthquake Policy" arrived in the mail from the California Earthquake Authority (CEA), emblazoned in boldface capital letters on the front was the following "disclosure":

> If losses as a result of an earthquake or a series of earthquakes exceed the available resources of the CEA, this policy is not covered by the California Insurance Guaranty Association. Therefore, the California Insurance Guaranty Association will not pay your claims or protect your assests if the CEA becomes insolvent and is unable to make payments as promised.

There was more. A surcharge of 20 percent of the premium, $60 in our case but much more for others, could also be leveled "when an earthquake or series of earthquakes has exceeded available resources to pay claims." Surely, I thought, these limiting provisions would become operable in a magnitude 8. It was not a comforting thought.

For other homeowners in higher-risk zones and older homes, there were substantial cost increases coupled with reductions in coverage. Consumers howled; public hearings were scheduled; there may be changes.

What were the experts doing about insurance?

A seismologist working in the Menlo Park USGS office said, "Around work there are plenty of people that both carry and do not carry earthquake insurance, but these differences are based on the various financial factors, as no one doubts that damaging earthquakes will continue to occur."

Lloyd S. Cluff was a longtime member and chairman of the California Seismic Safety Commission. He was also manager of PG&E's Geosciences Department. As such he was responsible for the seismic assessments of all the utility company's facilities, including the Diablo Canyon nuclear power plant. Cluff has done consulting work in 110 countries around the world.

The geologist lives in a large, older San Francisco home built by the person who was the engineer for the Golden Gate Bridge. The house sits on bedrock and is bolted to its foundation. The yearly earthquake insurance premium would be $7,000. The deductible would be enormous. Cluff carries no quake insurance.

Caltech's Hiroo Kanamori, professor of geophysics and the director of the seismology laboratory, is the holder of many prestigious awards. He is said to be the most brilliant seismologist alive today. When I talked with him in his office, Kanamori mentioned in an aside to a question on another matter that he carried no insurance.

He added, "My personal way of dealing with earthquakes is that I want to reduce the worry I have. I live in a relatively inexpensive home, a cheap house, really, so that the loss will be limited. I am very happy about it." Kanamori's forty-year-old home was located two miles from the Pasadena campus. The house had not been retrofitted. Perhaps the property was worth far more than the structure, as was the case with many older homes in California.

It was clear to me that earthquake insurance was not a given among the cognoscenti. There were, however, a number of other protective measures to consider.

Most likely we will have to be self-reliant for a number of days after a major quake. Flashlights, extra batteries, a butane lamp and extra fuel cartridges, candles, first aid supplies, a battery-powered radio, and camping food are stashed in our house. Next to my side of the bed is an extra set of dentures and a flashlight. In the back of my pickup truck, parked twenty feet from the house, there is a periodically updated copy of everything that is on my computer hard disk.

Seven gallons of water stored in two plastic containers sit in a storage shed next to the house, along with a camping stove, extra fuel, and more nonperishable food. There is additional camping equipment in a self-storage facility in town, which is one mile distant. I am thinking of bringing selected items home, since the two sliding doors that I need to open to gain access to the cubicle might jam in a quake.

There are some things that I haven't done. I have a pretty good idea where to turn off the water, electricity, and propane gas; but I need to ascertain where these valves are located and show them to my wife. I should investigate whether we need a special earthquake valve for the propane tank. We have to tie down, glue, or put nonslip mats under loose items, such as electronic equipment and dishes. I have a catalog for a company that sells such items but have not yet ordered any.

There is a neighborhood disaster council that is hooked in to an umbrella organization for all the small Tomales Bay communities. I have gone to one meeting, but I really don't know its current plan or recall my particular assignment.

My guess is that we are in the top 5 percent of California residents in terms of earthquake preparedness, but we have not done enough and will never be wholly prepared. It is, in the end, a matter of individual responsibility.

Just southeast of our home, the fault crosses Sir Francis Drake Boulevard and bisects a portion of Point Reyes National Seashore where there is a popular earthquake trail. The interpretive signs along the trail wrongly identify that spot as being the location of the epicenter and greatest movement in 1906. The fault then crosses the property of the Vedanta Society retreat in Olema, once the Shafter Ranch and the location of the cow episode.

Religion, like fresh water, bubbles to the surface along the San Andreas Fault in surprising amounts and varied forms. From Big Pines to

Pine Mountain in southern California there are rustic religious retreats of almost every order, including Saint Andrew's Abbey. The Spanish mission at San Juan Bautista sits adjacent to the fault in central California. Vedanta is an ancient religion that originated in India and is now practiced in northern California in a farmhouse of New England design. Overlooking the South Fork of the Gualala River is the glittering, copper-sheathed Odiyan Tibetan Buddhist Temple, built in the form of a three-dimensional mandala.

After Kerry Sieh's work in southern California drew a lot of attention, paleoseismic trenching became a fad. Trenches were dug in nearby Dogtown and on the property of the Vedanta Society. Tina Niemi, a graduate student at Stanford University, found evidence of up to five prehistoric earthquakes at Olema. When she combined her findings with those of other paleoseismologists who had worked in northern California, Niemi concluded that there was a large quake—termed a penultimate event—along the northern portion of the San Andreas Fault in the mid-seventeenth century.

My friends Richard Kirschman and Doris Ober live just to the south in Dogtown. The twenty-minute drive along the fault line is through some of the most idyllic countryside in northern California, a region known for its surface pleasantness.

The passage through fault-formed Olema Valley, an extension of Tomales Bay, is a study in contrasts: round golden hills to the east, dark Douglas fir forests on the ridge to the west. Along the swale between the two tectonic plates are thick oak trees, grazing livestock, wood and wire fences, and federal parklands. I always drive slowly through this idyllic pastoral zone.

There is a gate to open and close that keeps my friends' llama and exotic breeds of goats and sheep from escaping. A swatch of green marks where water has made its way to the surface through the fractured earth. The narrow, twisting drive leads to a dramatically sited home with a commanding view of the fault.

In 1906, a row of eucalyptus trees on what would become my friends' property was offset thirteen feet. Gilbert wrote: "The row is now both distorted and curved, and as there is reason to believe it was originally aligned with care, its present condition shows the distortion of the ground at the time of the earthquake." Richard has shown me the offset, visited on occasion by geology classes.

One night I arrived at my friends' home for dinner. The company represented a seismic cross section of the world's population. The host and hostess lived on the fault line. Another couple, Catherine Caufield and Philip Williams, were located on the Pacific Plate. My wife, who was absent that night, and I lived on the North American Plate. Annet and Jean-Francis Held of Paris were midplate residents and observers from afar.

After dinner we adjourned to the living room. I gave Richard a tape of earthquake sounds that I had purchased from the Seismological Society of America, and he popped it into a tape recorder that was hooked up to powerful speakers.

There were three types of noises on the tape compiled by Karl V. Steinbrugge, a University of California seismologist. There was the sound of the earth moving, described most frequently as low rumbling or distant thunder. Then there were the noises of structures cracking and breaking. Lastly there were the human sounds of screams and excited voices.

We listened intently. The human reactions were particularly riveting. There was a similarity: life was proceeding normally—a high school student was practicing the tuba, a lawyer was taking a deposition—then the sane world dissolved into a chaotic mess.

I asked the host to stop the tape after the ninth segment, a salsa tune recorded in a Caracas, Venezuela, studio by a musical group. Along with the technicians, the group had fled the studio and left the equipment running. The deep bass rumblings of the quake were superimposed over the treble tones of the bouncy tunes, a mix of the sublime and the ridiculous.

Jean-Francis commented, "But isn't this how life on earth will end?" He had touched on the apocalyptic nature of earthquakes.

A lively conversation ensued on the fault line.

Someone asked if there were videos showing the actual movement. I replied that the most common are from surveillance cameras that record the drunken movements of customers in convenience stores and the wild dance of merchandise on the shelves.

What about street sounds? The clamor of car theft alarms is the most prevalent urban noise.

The conversation turned to personal experiences.

"I love earthquakes," said Richard. Doris added, "In the middle of the night when a quake strikes I'm just horror-struck and clutching, and he's laughing and whooping."

Philip had just returned from Tokyo. He was in the fifth-floor conference room of a high-rise hotel when a small quake struck. After the short period of shaking and the dislodgment of a water tank on top of the hotel, the conference resumed.

We asked the French couple what their friends said when they announced they were going to California for their vacation. "Oh, they mention the earthquakes. They all saw it in 1989 on the television. Very dramatic," said Annet.

Our host recalled visiting in Holland some years ago. "Behind the house was a dike holding back the force of the whole North Sea," Richard said, "and this person wanted to know how the hell we dealt with earthquakes?"

We laughed heartily.

Catherine said she thought more people were killed in hurricanes. Her husband, an expert on rivers, said property damage from recent Mississippi floods was actually greater. Tornadoes were also discussed. We thought that of all natural disasters earthquakes were feared the most.

"I love the idea that we live on this plate and our friends are on the other plate," said Catherine.

She and her husband were thinking of building a new home. I pointed out that their septic system would be less costly than ours because the decomposed granite on the Pacific Plate was more porous than the clay soil on our side.

Richard suggested, "Instead of saying we are going 'over the hill' [the local expression for driving over the Coast Range to urban areas], why not say we are going 'across the plate'?" Someone said they needed to cross the fault, and soon thereafter we departed.

A short distance to the south the fault dips into the ocean at the point where an inlet separates the iconoclastic community of Bolinas from the more conventional beach resort of Stinson Beach. Every Fourth of July there is a tug-of-war match between the two villages. The losers are dragged into the inlet. A parade down Wharf Road in Bolinas, where structures were tipped into the lagoon in 1906, follows the physical contest that mimics the movement of the land.

TO THE SOUTH

When the shock wave struck the San Mateo County shoreline just south of the epicenter, it shook loose a landslide some two hundred feet wide and fifty feet deep that buried the beach at Mussel Rock. The landslide hid the newest fracture line of the San Andreas Fault; but the scarred, cavernous bluff was evidence enough of previous faulting. The fissure surfaced near a wagon road in what was then open farmland. It spread to the southeast toward San Andreas Lake, part of San Francisco's water supply.

Three dams in the reservoir system were on or near the fault line. The first two were earth-fill structures. San Andreas and Upper Crystal Springs Dams were directly on the fault. The violent movement offset both dams eight feet. Their banks cracked but the dams held. Lower Crystal Springs Dam was constructed of concrete blocks. It paralleled the fault and lay about one thousand feet to the east. There were fissures in the earth around the dam, but no cracks were found in the structure.

It was not for lack of stored water in the reservoirs outside the city that there was insufficient water to fight the resulting fire in San Francisco. Rather, breaks in the distribution system within the city cut the flow of water. Charles Derleth Jr., the structural engineer from the University of California, surveyed the city waterworks after the quake. He concluded, "For safety the city needs a number of sources of water, in localities widely separated and not in the same geological region, with a number of main conduits so arranged that they will not tend to be destroyed all at the same time."

Farther southeast at Stanford University, fifteen years old and off to an ostentatious start with millions of dollars of railroad money, the damage was extensive. Among other buildings destroyed was the new geology building, which had not yet been occupied. A student was killed in a dormitory. Fortunately it was too early in the morning for classes; otherwise the death toll would have been far greater. Three brick chimneys and the plaster collapsed in President Jordan's house.

The inmates at nearby Agnew's Insane Asylum were not as fortunate. More than one hundred patients and staff were killed. Derleth noted: "Of all the destruction that I saw, and I visited the whole disturbed area, this cluster of buildings exhibited the most complete earthquake destruction, with the possible exception of the City Hall buildings in San Francisco. They are both public structures. Is it not time for California to realize the situation?"

The USGS report on structural damage, of which Gilbert was a coauthor, was more scathing. The insane asylum's buildings were "all flimsily constructed brick structures with timber frames. The construction of these buildings, with their thin walls (in many places devoid of mortar) and light, insufficient wooden framing, indicates a criminal negligence that is appalling," stated the report.

Between one thousand and eleven hundred people were at Agnew's at the time of the earthquake. No one knew the precise number of dead, as the records were buried with the unrecovered bodies. Perhaps twice as many were badly injured. A reporter for the *San Jose Herald* noted: "A large body of men is at work delving into the ruins in search of bodies. But the tragic scenes that constantly pass before their eyes and those of the heroic attendants have awakened in their hearts a pity for the sufferers."

The sister of the county coroner fell from the fourth floor of the Administration building to the basement, where she was buried for many hours. One hand was free, and by waving it she attracted the attention of rescuers. Four dead people lay around her. The bodies of the dead overflowed the morgue and all the local funeral parlors.

The coroner immediately scheduled an inquest into the death of one of the asylum's supervisors that covered all deaths. Death by shock from the earthquake was the jury's hasty determination, adding "that we find that many other deaths occurred in Santa Clara County from the same cause."

There were fears that the violent inmates would get loose and roam about the countryside. Some were shipped by train to the Stockton asylum. Armed deputy sheriffs patrolled the area. On the asylum's grounds there were not enough tents for all the dazed patients. For a while there was no food. "The violently insane have been confined in an enclosure where their cries add to the terror that still envelops the place," wrote the reporter four days after the earthquake.

Despite these deaths and the destruction of almost all brick and stone buildings in downtown San Jose—an amount of devastation termed "appalling" by Derleth—the *Herald* found "a brighter side." In an editorial by that title, the publisher stated that since no earthquake, except volcanic shocks, had ever struck twice in the same place, "San Jose is today safer from danger from earthquakes than ever before." His source was a one-day "investigation of all available authorities on the subject." He continued:

> This carefully recorded history of these disturbances therefore gives us every assurance that San Jose, San Francisco and the region now visited by this temblor is with reasonable certainty exempt for all time from a recurrence of this experience.

Of the area now known as Silicon Valley, the publisher went on to prophesy, if not command:

> SAN JOSE WILL BE REBUILT. It will be made more substantial and more beautiful than ever. We still have our sunny skies, our invigorating breezes and our fertile soil.

TO THE EAST

To the east of the epicenter, the earthquake and resulting fire that razed vast portions of San Francisco was the greatest natural disaster to ever befall a North American city. It is an oft-told tale that has been muted over the years. Time, of course, has lessened the immediacy, as has the human manipulation of the historical record.

The 1906 earthquake was the first to be extensively documented in black-and-white still photographs. Color photography, which came later, would have been superfluous. The scenes were essentially grim and colorless.

Two aspects of the photographs are striking. For a country untouched by the destructive power of modern warfare on the scale of a Dresden, London, or Hiroshima the amount of devastation was beyond belief. It greatly exceeded the fatalities and damage from wildfires, floods, and urban riots in disaster-prone California for the remainder of the present century.

In the midst of the smoking ruins the people—be they refugees, rescuers, or onlookers—were all well dressed: two- and three-piece suits, white shirts and neckties, and derby hats for the men; long dresses, shawls or jackets, and large, floppy hats for the women. The comparable sartorial scene today would be San Francisco's financial district at noontime.

The dichotomy between the scale of damage and the mode of dress was striking. It was a different time with different values. People—whether in Santa Rosa, San Jose, or San Francisco—dressed for disaster. They also may have escaped with only their best clothes. A contemporary observer noted with pride: "The San Francisco man wore laced boots while the debris was being cleared away; the San Francisco woman never laid aside her kids and suedes and ties for a day." (Earthquake fashions change. In 1989 they wore sweatsuits, Levi's, shorts, and T-shirts.)

Words accompanied the images. California's finest writers were in the Bay Area at the time and wrote accounts of the tragedy. Mary Austin was living in San Francisco; Gertrude Atherton was in Berkeley and traveled by ferry to the city; and Jack London rushed to San Francisco from his ranch in Glen Ellen, stopping by Santa Rosa on the way.

The beginnings of a literature of disaster in California were not auspicious. Austin's account, "The Temblor," published in *Out West*, was mannered; Atherton's story in *Harper's Weekly* was distant; and London admitted he was overwhelmed by the immensity of the event. However, one quote from London's short article in *Collier's* magazine does bear repeating. He wrote, "All the cunning adjustments of a twentieth century city have been smashed by the earthquake." London cited James Hopper's account in *Everybody's Magazine* as the best quake story.* Hopper also wrote an article for *Harper's*.

Hopper was a reporter for the *Call*, Mark Twain's old paper. He went on to write magazine articles and books and secured a minor niche in California's literary history. For the story of the earthquake he was the right man in the right place at the right time. Starting with waking up and watching the outside wall of his downtown apartment disintegrate, Hopper seemed to be everywhere. Unlike the others, who carried their societal and political baggage with them, the reporter was absorbed by the immediate scene.

In his *Harper's* story, Hopper accurately described the incredible demands and the magnetic attraction of disaster reporting. Hopper and other reporters commandeered an auto. He wrote:

> It was a phantasmagoria of destruction. We ate a sausage here, a
> cracker there: we wrote upon our knees in haste, throwing the copy into

* I disagree. Hopper's account in *Everybody's* is forced—the consequence of having to come up with additional material after writing an earlier, more powerful account for *Harper's*.

a launch impatient for the presses across the bay; we rescued, carried wounded, helped to vacate burning hospitals; but through it all we circled that fire, circled and circled it as if fascinated. . . .

The reporter depicted the dazed state of mind that exists during such calamities:

> That diabolical earthquake had given us such a shake, that long minute had been such mental torture, that our brains were numb. We did not realize the extent of what happened and was happening, and we were never to do so. The disaster was one long, three days progression; by the time one phase of it was grasped it had swept on to another, and when it was all over the entirety was so colossal as to be beyond the immediate realization of human minds. The destruction of San Francisco will always remain a vague, chaotic, and somber nightmare.

From an overview he circled in to describe details, in this case fleeing refugees and their towering nemesis:

> There is no panic, no jostling, no running, no trampling. They simply march, heavy-stepped, and somehow the very calm of it is far worse than the hysteria of panic. It tells of greater tragedy, of more complete hopelessness. The faces are of stone, the eyes are dead, there is no revolt; and behind, its advance comber almost above them, the great tidal wave of fire.

More than anything else, he noticed the silence. He described it as "a heavy, brooding silence" and "a contortion of stone, smoke of destruction, and a great silence."

The editor of *Harper's* ran pages of earthquake photographs and let them seamlessly merge with photos of the recent eruption of Vesuvius. The implication was that such destruction, whether it be an Italian town drowned in lava or a scorched American city, had a commonality and was not limited to a single time or place.

Statistical facts followed the initial flow of descriptive words. The one minute of violent shaking caused the firestorm that raged for three days and two nights in San Francisco. The earthquake, which struck without warning, was the cause of most deaths. People could walk away from the fire, for there was ample warning of its approach. The fire caused most of the damage. Generating its own winds and temperatures that reached 2,700 degrees Fahrenheit, the firestorm consumed the heart of the city.

The casualties and destruction exceeded previous great fires in London, Chicago, and Baltimore. Three thousand acres, more than four square miles, or a little over five hundred city blocks were destroyed, leaving but a few outlying residential areas intact. Twenty-eight thousand buildings were ruined. The army and city officials put the number of deaths at between five hundred and seven hundred, a figure that would be revised upward by a factor of five some eighty years later. Along with deaths in the Chinese and Latino populations, the maimed and injured seem not to have been counted.

Immediately following the quake the scientific community went looking for causes and, like everyone else, silver linings.

Stanford's Jordan ruled out electricity, planetary conjunction, earthquake weather, and "the wickedness of man" theories—all of which had been advanced by others. The cause was strictly mechanical. The break was induced by the vertical rising of the coast, Jordan said, a theory advanced by Andrew C. Lawson of Berkeley in the previous decade. But Lawson was now focusing on horizontal movement.

Benefits were cited. "Fertile and well-watered" valleys, "often marked in California by successions of dairies and reservoirs," were one of the advantages of extensive faulting, said Jordan. Harold Fairbanks, the USGS geologist who had traced the San Andreas Fault before the massive break, philosophized: "Although we dread earthquakes with all their resultant destruction, yet it is well to recognize the fact that if it were not for them we would find here in California little of that wonderful scenery of which we are so proud."

Three days after the quake Frederick Leslie Ransome of the USGS sat down in Washington, D.C., to write an article for the National Geographic magazine. Ransome thought the city would be rebuilt in a safer manner, as did his colleague Gilbert on the opposite coast. Both scientists were naive.

WHAT QUAKE?

San Francisco needed to rebuild and rebuild quickly. I don't argue that point. But it should have rebuilt intelligently, employing the lessons of previous quakes. It should not have effaced the nightmare of those days and nights. Unfortunately, that was the California way.

Instead of toughening laws, building codes were actually relaxed in order to hasten rebuilding. In fact, a weakened building code was adopted in May, before engineers had completed their studies of earthquake damage and forwarded their recommendations for changes in the code. The opposition of the bricklayers' union prevented widespread use of reinforced concrete, regarded at the time as the most earthquake- and fire-resistant of all building materials.

A USGS study of building practices and materials cited cost cutting and "dishonest design and construction" as the main reasons for the failure of structures, and concluded: "It is very probable that the new San Francisco to rise on the ruins will be, to a large extent, a duplicate of the former city in defects of construction."

Actually, the gleaming new city was structurally more unsound than the old city and would remain so, at least until minimal seismic design requirements were adopted after much debate in 1948. Twenty years after the quake a noted engineer told a Commonwealth Club audience: "With respect to safe design . . . we have actually seriously retrogressed."

Laws and ordinances, when finally enacted, did not necessarily alter the situation. Two engineers writing in a 1980 California Division of Mines and Geology special report on the northern California portion of the San Andreas Fault noted:

> The Uniform Building Code has had various seismic provisions over the last several decades, but these have not been adopted or enforced uniformly, promptly, or consistently over the years by the various local governments.*

Three years after the 1906 quake twenty thousand new buildings were in place. The process of rebuilding created jobs, depleted the forests, and took the lives of fifteen thousand gaunt horses who were worked to death. The debris was dumped in Mission Bay and the Marina District, creating more made land—exactly the type of terrain that is most vulnerable to shaking.

---

* Seven years after the 1989 Loma Prieta quake, work had barely started on retrofitting eighty-nine high-risk brick structures in San Francisco. One such structure killed five people in that quake.

The rewriting of history commenced shortly after the ruins began to cool. The dual catastrophes became one, known as "the Great Fire." The fire was emphasized for two reasons. Earthquakes were unpredictable affairs that made financial markets back east very nervous. Eastern money was needed to rebuild San Francisco. Most insurance policies did not pay off for earthquakes, but they did cover fires.

The revision of history was led by the Southern Pacific Company. What the state's largest business achieved, said Gladys Hansen, coauthor of *Denial of Disaster*, "was, and is, breathtaking."

The railroad, through its glossy publication, *Sunset* magazine, set about sanitizing the disaster in its May issue. The magazine employed the best available illustrators, photographers, writers (London and Austin, among others), and editors. It had four hundred thousand readers.

The image the magazine projected of California and the West—one that was nurtured by succeeding owners, including Time Warner at present—never matched the gritty reality of the region. But it certainly encouraged its readers to take those idyllic vacations at Lake Tahoe, build those tasteful redwood decks, plant those lush gardens, serve that fresh food, and support the advertisers who sold all of the above plus more of the same kind of safe western fare.

The earthquake threatened this carefully crafted image of pastoral beauty, so the *e* word and all it stood for had to go. The planned May issue was destroyed as it sat on the presses. A brief "emergency edition" replaced it. The black-and-white cover was an illustration by Maynard Dixon, who would emerge as a leading painter of Southwest scenery. Dixon drew "the Spirit of City" in the form of a nude, nippleless woman rising phoenixlike from the burning buildings.

By the time of the June-July issue, when the format had returned to normal, the ink drawing had been altered and was now rendered in the strong, bold colors for which Dixon would become regionally famous. The woman's breasts were now demurely hidden behind the gleaming new city she embraced as she rose from the smokeless ruins. Accompanying the illustration was a poem by Dixon entitled "San Francisco Promise."

By the October issue, new steel girders in the foreground framed an exaggerated amount of completed construction in the background of the cover illustration, this time rendered by a different artist. The articles were entitled "San Francisco Upraising," "Some Reconstruction Figures," "San Francisco at Play." The author of the last story complained that "outsiders are coming to think that the city is in a sackcloth as well as in ashes. Nothing could be further from the truth." San Francisco, he said, "is still the gayest city in the Western Hemisphere." The magazine avoided articles about poor or nonexistent sanitation, the spread of deadly diseases, mental illnesses, suicides, and increased use of cocaine and opium.

With the most to lose and the most to gain, Southern Pacific continued to push the recovery and the denial of the earthquake, and others fell into line. Rufus Steele, a socially well connected writer whose only other effort of note was a play, *The Fall of Ug*, written for the annual Bohemian Grove encampment, wrote an article for the April 1909 issue of *Sunset* titled "The City That Is."

The title alluded to a popular booklet written by Will Irwin of the *New York Sun*. The booklet was an expanded version of his story in the April 21 edition of the New York newspaper. The story, titled "The City That Was," began: "The old San Francisco is dead."

After the dramatic beginning, Irwin's quickie book settled down to a rather maudlin, highly romanticized re-creation of the former city, as recalled by a facile writer. It ended: "The bonny, merry city—the good, gray city—O that one who has mingled the wine of her bounding life with the wine of his youth should live to write the obituary of Old San Francisco!" The forty-seven-page paean was reprinted at least four times.

Later that year Steele published a book with the same title as his magazine story. He used the *e* word twice. Neither time was it associated with actual destruction. (As a fellow writer, I was fascinated by how Steele could evade mentioning the cause of the fire. Easy. He avoided referring to the events in the early-morning hours of April 18 by starting his account of the devastation in the following manner: "On the night of April 20, 1906, after three days of burning, burning, burning, the city lay in shards.")

Newspapers were part of the chorus of denial. Arnold J. Meltsner, a Berkeley political scientist, later pointed out:

> The simplest way for a newspaper to deal with an earthquake was to refuse to acknowledge its occurrence. A good example of this was the reference to an earthquake in San Francisco in 1906 as "fire" or as a

"conflagration." Of course, in something as tangible as and visible as an earthquake, it is practically impossible to disregard it altogether. It can, however, be interpreted as something else by omitting any reference to certain of its causes or consequences and by focusing attention on events which are incidental to it.

The *San Francisco Chronicle* printed a long story about the "more severe" Charleston earthquake. The writer of the first-person account was a child in Charleston at the time. The headline contained the moral of the story: "Lesson Learned From the Charleston Quake, How the Southern City Was Rebuilt Finer Than Ever Within Four Years." The mantra was buried in the article: "It was not the earthquake but the fire that wrought the destruction in this great city."

All the slanted prose and official proclamations failed to hide the reality from just those people it was intended to impress. Actually, the propaganda worked against the interests of San Franciscans, who may have wound up deluding only themselves.

In October of 1906, a civil engineer wrote a letter to Gilbert stating: "The people have already succeeded in psychologizing themselves to the belief that it was a fire rather than an earthquake—all except the insurance underwriters."

John R. Freeman, a Rhode Island engineer and president of a large insurance company, was considered the outstanding authority on earthquakes and insurance. Freeman wrote in 1932: "There seems to have been a local attempt to suppress information about earthquakes, lest it hurt California business, and rumors of this suppression have reacted unfavorably, by increasing the apprehension of Eastern underwriters."

Along with the disinformation, lack of interest combined with political pressure to blunt the spread of information about seismic dangers. Fred G. Plummer, a Washington, D.C., civil engineer who worked extensively in California, sent Gilbert a private communication. Gilbert had asked Plummer to write a report on his observations of the quake. Plummer, who was in a San Francisco hotel at the time of the temblor, did so; and his observations also covered the state of mind of Californians.

Many, both in and outside the state, Plummer wrote, were ignorant of previous temblors. Plummer could not locate a recent catalog of California quakes in three Bay Area libraries. On his return to Washington, he went to the USGS library and found the 1898 Smithsonian Institution report. The pages were uncut. Plummer added, "Several teachers of geology in the West have informed me that they have never seen the catalogue although it was issued in 1898 and was reprinted for distribution by the Government."

He was referring to A *Catalogue of Earthquakes on the Pacific Coast 1769–1897*, compiled by E. S. Holden, director of the University of California's Lick Observatory. Seventy years later a USGS publication termed the catalog "a cornerstone" for all such future compilations.

Plummer noticed that scientists were hesitant to speak out on the dangers of earthquakes. "Moreover, there seems to be some willful ignorance among teachers of science in the West on this matter, due in part to the fact that their views, if expressed, would not be well received."

Not only the West, but the country as a whole was woefully ignorant. There was no chair in seismology nor a department of seismology in any American university. A course in seismology, disguised as Geology 114, was introduced at the University of California in 1911 and most probably was the first that directly bore on the subject in this country.

The engineer continued, "Frankly, the people do not want to know the facts. It hurts real estate. Yet I would as soon live in San Francisco as elsewhere, so far as this risk is concerned, provided the full lesson of the temblor had been learned, which it has not."

Against this background of hostility and indifference, the California Earthquake Investigation Commission set out to compile a report that would provide an honest record and educate Californians about earthquakes. The commission was named by the governor three days after the temblor. Gilbert was a member, as was Branner of Stanford. Lawson of the University of California was its chairman.

The commission was soon urged to desist from its appointed task. Branner and Lawson, who had their personal differences, agreed on the obstacles.

Branner later wrote:

> Shortly after the earthquake of April 1906 there was a general disposition that almost amounted to concerted action for the purpose of suppressing all mention of that catastrophe. When efforts were made by a few geologists to interest people and enterprises in the collection of

information in regard to it, we were advised and even urged over and over again to gather no such information, and above all not to publish it. "Forget it," "the less said, the sooner mended," and "there hasn't been any earthquake," were the sentiments we heard on all sides.

Five years after the quake, Lawson made the following observation:

In the present state of public opinion in California for example, it is practically impossible to secure state aid for the study of earthquakes. The commercial spirit of the people fears any discussion of earthquakes for the same reason as it taboos any mention of an occurrence of the plague in the city of San Francisco. It believes that such discussion will advertise California as an earthquake region and so hurt business.

Gilbert summed up the post-1906 mood in his address to the American Association of Geographers. He said:

When the Seismological Society was under consideration, there were businessmen who discouraged the idea, because it would give undesirable publicity to the subject of earthquakes. Pains are taken to speak of the disaster of 1906 as a conflagration, and so far as possible the fact is ignored that the conflagration was caused, and its extinguishment prevented, by injuries due to the earthquake. During the period of aftershocks, it was the common practice of the San Francisco dailies to publish telegraphic accounts of small tremors perceived in the eastern part of the United States, but omit mention of stronger shocks in the city itself; and I was soberly informed by a resident of the city that the greater number of the shocks at that time were occasioned by explosions of dynamite in the neighborhood. The desire to ignore the earthquake danger has not altogether prevented the legitimate influence of the catastrophe on building practises, but there can be little question that it has encouraged unwise construction, not only in San Francisco but in other parts of the malloseismic district.

Considering the obstacles, it was a miracle that the commission produced any report, let alone the most comprehensive, literate, and understandable such document ever published on the subject.*

---

* The two-volume report, paid for and published by the Carnegie Institution of Washington, D.C., is still in print.

What the commission eventually accomplished in 1908 put all such past and future efforts to shame. Instead of the esoteric technical jargon and narrowness of viewpoint that plagues most contemporary efforts, the work—written in plain English and handsomely illustrated—was and is universally praised for its clarity and content.

Perhaps because the commissioners had limited seismic experience, the report differed in "freshness and freedom" from the usual effort. Those were the words of Charles Davison, a noted English seismologist. Writing in the centennial volume of the Geological Society of America, which dealt with the history of North American geology, University of California seismologist and author Bruce Bolt stated: "This treatise remains a model of an effective study of a great earthquake, and it should be required reading for all those interested in what is likely to happen in the next great California earthquake."

Lawson deserved much of the credit for the timeless document. He took a leave as chairman of the geology department but still took care of administrative details, like budgets. Surely, he thought, the university would increase seismic research after the quake. But that was not to be the case. Lawson said there was "little to be hoped for" from President Benjamin Ide Wheeler.

Lawson's personality was characterized as vitriolic by Charles Richter of Caltech and others, but that doesn't show up in his voluminous correspondence with commission members or other scientists who participated in the study. He was firm, he cajoled, and he pleaded that they get their work done on time. Then he probed for soft spots and inconsistencies, seeking thoroughness and accuracy. Lawson also paid a great deal of attention to the maps and photographs and their placement in relation to the text.

He depended to a great extent on Gilbert, and the two geologists batted ideas back and forth. It took a long time for the two men to hammer out the definitions for *rift* and *trace* and whether there should be a composite text or separate contributions. Lawson decided on the latter solution.

About half the commissioners were astronomers. When Harry Fielding Reid of Johns Hopkins University arrived, Lawson wrote Gilbert, "I am glad he is here, as I have been feeling geologically lonely on the commission."

Branner was another case. Their relations were testy. Lawson kept after Branner to complete his description of the faulting between San Francisco

and San Juan Bautista. "I know you are fully occupied with other matters
and have little time for getting this earthquake work out, but . . ." When
that didn't work, Lawson asked that "your young men" do the job. Branner
assigned graduate students to the work, and then left for Brazil.

The rivalries of competing academic institutions were very much a part of
the post-earthquake scene. Branner found time to contribute an article to a
popular book compiled by Jordan. This book, mostly a Stanford effort, beat
the commission's report, viewed as a University of California–dominated
work, into print by one year. The great esteem in which Gilbert was held can
be judged by the fact that he contributed to Jordan and Lawson's efforts and
the USGS report.

Lawson first estimated that the report would take six months to complete. *The California Earthquake of April 18, 1906* was published in December of 1908 at a total cost $5,640. A second volume, *The Mechanics of the
Earthquake*, containing Reid's elastic rebound theory, followed in 1910. The
first extensive description of the San Andreas Fault, the thorough documen-
tation of earthquake damage, Reid's theory that persists to the present, and
Gilbert's details of surface faulting in Marin County were the report's most
noteworthy achievements.

Unfortunately, San Franciscans had gone on to other matters by the time
the studies were completed, and they barely caused a ripple when they were
finally released. The city was focused on planning for the stupendous
1915 Panama-Pacific International Exposition, which would provide visual
proof—by its very opulence staged on made ground in the Marina District—
that there had never been a devastating earthquake.

Along with the cleansing effect of an exposition and the rewriting of history, cartography was also altered to prove there had been no earthquake. An
atlas with twenty-five maps that clearly delineated seismic fissures throughout the state accompanied the 1908 report. Eight years later the trouble-
some faults had vanished. The "Geological Map of the State of California"
was published in 1916 by the State Mining Bureau. The map, compiled by
J. P. Smith, a professor of paleontology at Stanford University, was devoid of
any fault lines. The oversight, termed "puzzling" by later mapmakers, was
not rectified until 1938, when faults were first shown on a map published by
a California agency. The most recent such map, published in 1994, shows
more than five hundred faults.

Seismologists through the middle years of the present century continued to be aware that the word *earthquake* had disappeared from the vocabulary used to describe the events of 1906. However, they lacked the interest and the will to go public in an effective manner to remedy the distortion. It took an archivist in the city library of San Francisco to uncover the propaganda effort and come up with a more realistic casualty figure.

Gladys Hansen was in charge of the genealogy collection when the San Francisco Room opened in the library in 1963. Almost every day someone asked: "Do you have a list of those who died in 1906?" Mrs. Hansen could find no such list, so she set out to create one. With the help of others she combed existing records and eventually came up with a figure of more than three thousand deaths in San Francisco alone.

The number is a minimum figure. It does not reflect deaths in San Jose, Santa Rosa, or other outlying areas; the work has not been completed for San Francisco; and the total just scratches the surface of the minority populations, such as the Chinese, who either took care of their own casualties or were not counted by officials.

Because of the very nature of the cataclysm and the human response to it there will never be a final determination of the number who died. But old habits are hard to break, especially when they have an "official" sanction. A current USGS "fact sheet" entitled "When Will the Next Great Quake Strike Northern California?" places the 1906 casualty figure at seven hundred.

7 PARKFIELD
(1966)

FROM HOLLISTER TO PARKFIELD

The one-hundred-mile section of the San Andreas Fault from near Hollister to just south of Parkfield in central California is the best place to observe the effects of gradual tectonic motion, otherwise known as *creep*. While other sections seem to be locked in place, until they periodically explode in violent jerks, the central section moves in tiny increments.

There is a neatness of concept, as well as movement, to this section of the fault line. Both Hollister, a small city and county seat, and the hamlet of Parkfield have laid claim to being the earthquake capitals of the world. These assertions were a way to attract tourist dollars. Hollister, more interested now in becoming a suburb of Silicon Valley, has quietly dropped the moniker. Since the Parkfield prediction experiment failed, the ranching community's claim to fame rings hollow.

There are still good tectonic reasons to visit both communities. Except for the offsets in the Carrizo Plain, the bending—sometimes to the point of breakage—of human artifacts that cross the fault line is the most visible manifestation of the presence of great underground forces at work.

The inexorable movement of the two plates is manifested by the bent curbs, sidewalks, and retaining walls of Hollister; the torn floor and dislocated beams and walls of a winery in nearby Cienega; and the bow-shaped bridge at Parkfield. It is at these places that the eventual failure of our constructs, if not constantly maintained, is foreshadowed.

Picture a weakened civilization and imagine how its physical deterioration would be hastened by the constant tearing apart of its underpinnings. I think this is a much more realistic scenario of apocalypse than any of the instant quakes and fires that flash across the surfaces of screens, pages of text, or religious imaginations.

Whereas a visit to the Carrizo Plain requires special preparations, these destinations are reachable by paved roads. However, the fault line between Hollister and Parkfield is relatively remote, and the interior valleys of the Coast Range are quite hot during the summer and early-fall months.

For my journey I chose mid-March, the height of the wildflower season. The green grass on the rounded hills and mountains, dotted with lollipop oak trees, rippled in the spring winds. There was a sense of freshness in the perfumed air. Because of its rural serenity, this is one of the most beautiful parts of California.

You never know what lies underneath you in this state.

While researching local earthquakes in the San Benito County Free Library in Hollister, a librarian pointed to the line of slightly darker, obviously newer linoleum tiles laid in a six-inch swath across the floor. She said, "This is our notorious earthquake souvenir. They have to be replaced every once in a while, especially after the bigger shocks." The single line of tiles passed directly underneath the filing cabinet containing newspaper clippings of past earthquakes.

Hollister is located on a shaky foundation. The city is underlain by alluvium, the result of deposits from the San Benito River over time. Lawson's report on the 1906 earthquake stated: "One old settler remembers when the business part of Hollister was a slough. An artesian belt also passes thru the town, which may have affected the intensity along its path." Locating a small town here guaranteed its future designation as the earthquake capital, since Hollister was near the junction of the San Andreas and Calaveras Faults.

In 1906, three brick buildings in the downtown section collapsed, and two people were killed. The structures were poorly built—large rooms with no interior supports. The firehouse and a school were badly damaged, but children were not in classes at that early hour. Half the chimneys in town collapsed. Residents first heard a rumble and then were hit by two shocks a few seconds apart.

Eighty-three years later 170 residences and more than 20 commercial structures were damaged in the 1989 Loma Prieta quake. Four people were seriously injured. Underneath a poster for the movie *Mississippi Burning* in the window of a local video store a wag had placed the hand-lettered sign "and Hollister Falling."

An average of eight earthquakes per year have been felt in Hollister since 1928. Only a few caused hardship. Any damage is quickly cleaned up, structures are repaired or new ones built, and within a few years the memory of the last bad quake is conveniently erased. What is more noticeable is the creep along the Calaveras Fault through the older section of town.

Rather than constant, the surface movement is episodic and is attributed to underground faulting. Thomas H. Rogers of the California Division of Mines and Geology wrote: "Commonly, but not always, fault slip episodes are followed (in 2–5 weeks) by surface fault creep events, or by earthquakes, or both."

I stopped at the chamber of commerce office on San Benito Street and asked for information on earthquakes. I was handed a poor-quality photocopy of a guide to the fault prepared by Rogers some years ago at the request of the city. I was also handed a recent color brochure titled "San Benito County: California's Unspoiled Paradise."

There were two references to earthquakes in the slick brochure. The history section mentioned that the county "was one of the most active earthquake areas in the world." Under the category of "Parks & Recreation" was the listing for the San Andreas Brewery Company: "Stop by to view the fermentation tanks and order a plate of nachos rated on a Richter scale of hotness and home of the famous earthquake burger."

I followed the directions in the fault guide to Sixth Street, just west of West Street. It was here that the slow tearing apart of the earth's surface was most apparent. Rogers noted "the sidewalks, curbs, and concrete pavement slabs are all offset across a wide zone." He measured a four-inch offset in the north curb that had accrued over forty-five years. There was a photo of the broken curb in his report.

The gap has noticeably widened in the last quarter century. Using rudimentary tools, I measured an eight- to nine-inch offset in the curb. What was quite striking was the much greater displacement in the adjacent sidewalk and retaining wall.

As I was trying to figure out what was what, a man came out of the small Victorian house that fronted on the broken and bent objects and asked me: "You figure out the difference between the nine inches in the curb and the twenty-three in the sidewalk and wall?"

Jerry Damm's home was built in 1878. He has retrofitted it for quakes: putting in a new foundation, strapping down the water heater, and doing other recommended repairs. There was no damage to his house in 1989. Two houses in the neighborhood slipped off their foundations.

Damm's next-door neighbor lost a lot of china: "He said, 'How'd you guys do?' I said, 'Fine. The Tupperware came through without a problem.' "

The Calaveras Fault cuts across Damm's property. The mystery, he said, was how the barn had stood for more than ninety years. "The barn is built on a mudsill, probably around the turn of the century. It's twenty-six feet wide, thirty-four feet long, and has no center support. There is a full hayloft. And it's still standing. I don't understand."

Damm departed for work, and I was left with the mysteries of the differing offsets and the resilient barn to ponder.

The phenomenon of creep tugs incessantly at the Cienega Winery, and for good reason—the structure was built directly on the moving fault line. Tectonic movement must be good for wine grapes, since the vineyard through which the fault line also passes is one of the oldest in the state. Periodically over the years, wine produced from these grapes has won medals at the Paris Exposition and the California State Fair, and it has been rated among the best in the nation.

The name of the winery was also a clue to the presence of seismicity. *Ciénaga* means "marsh" in Spanish, and there is a shallow sag pond just to the north of the winery, located seven miles south of Hollister on Cienega Road.

It was a building inspector who first took official notice of the cracked and displaced reinforced-concrete and wood-beamed winery in 1956, although previous owners must have known about the problem for at least one hundred years. A Frenchman first planted grapes in 1854. There was minor damage in 1906.

Shortly after the building inspector's visit, University of California seismologists arrived on the scene to measure and evaluate the new phenomenon. The Berkeley seismologists noted:

> There is local opinion to the effect that the previous building at this site had been damaged in a similar manner. The description of the damage (skewed roof trusses), and wall offsets of about two feet in perhaps half a century, suggest that the rate of creep of one-half inch per year may have been fairly constant for many years.

Near Cholame, at the opposite end of the moving section of the fault, the rate was one-tenth inch per year. The significance of the finding was that creep now had to be considered one of the mechanisms that relieved the accumulation of strain on faults.

Through a parade of recent owners—including such giants of the liquor industry as Hiram Walker, Almaden, and Heublein—the damage to the winery has been fairly consistent. Small quakes helped the process along.

A partnership took over the Cienega Winery in 1989. Pat DeRose, one of the partners, gave me a tour of the large gray shed in which there were stainless-steel vats, redwood barrels, and the sweet-sour smell of fermented grapes. DeRose described his duties as owner, winemaker, farmer, and repairman.

In his office, attached to the south side of the winery, there was a jagged gash in the linoleum-and-concrete floor of about one and one-half inches in width. Oriented in the general direction of the fault, the crack was the surface manifestation of the fissure that cleaved the earth's crust and underlying lithosphere.

Which way did the plates shift in his office? I asked. DeRose said, "No one knows. No one has ever been able to figure that out. It could go this way one day, then back that way, then this way."

Inside the winery, DeRose showed me various recent displacements, some measuring nearly two feet. I asked him if there were any sounds associated with creep. "Wood cracking," he answered.

Like Sisyphus's struggles, repairs were constant. DeRose pointed upward: "All those rafters were ready to snap. We ripped the whole ceiling out until you could see the sky. Put in all new rafters and new joists. New everything. Reinforced everything. So this thing is really, really set up. It is built! We spent fifty, sixty thousand dollars. Now we are ready to do some more work."

The partners bought the winery shortly before the Loma Prieta earthquake. There was a sharp jolt, but only two water lines snapped.

DeRose recalled a story he had been told about an earlier earthquake: "A nine-thousand-gallon tank toppled and took out a few of the barrels with it. The wine went running down the road to the pond, and all of the ducks got shitfaced. They found drunk ducks walking down the road. That is part of the history."

What do you do for earthquake insurance? "Pray," said DeRose. "Insurance companies aren't stupid, you know." He added that the premiums were prohibitively high.

Outside in the older part of the vineyard, the rows of vines were uneven. DeRose described what it was like driving a tractor: "It's really strange. I get to a spot in the vineyard, and I've got to go over two or three feet. What we've done is clear a path between the two sections so that I have time to cut my tractor over."

I knew there had been a plaque attached to the winery that designated the San Andreas Fault as a national historic landmark. I couldn't find it. DeRose said that a redwood tank weighing five tons had dropped from a forklift one day. It demolished the wall that held the plaque. He still had the bronze plaque and was waiting for some government agency to tell him what to do with it.

As I left, DeRose gave me a bottle of premium wine, a 1992 cabernet sauvignon with a rose on the label. "DeRose Family Vineyard, Cienega Valley," it read. I wondered why they hadn't added: "Crushed naturally by the San Andreas Fault."

The English navigator and explorer George Vancouver, in the company of a party from Monterey, visited the Pinnacle Rocks in 1794. He later wrote they were "the most extraordinary mountains I had ever beheld."

Vancouver knew mountains, having twice sailed around the world with Captain Cook. He was nearing the end of his own extensive voyage of discovery in 1794 and had spent a good deal of time in the Northwest and Alaska.

What Vancouver would have said if he had known that this volcanic formation had moved 190 miles to the northwest in the last twenty-three million to twenty-four million years can only be surmised. My guess is that he would have found a classical allusion for tectonic displacement, much as he did for a description of the volcanic monoliths.

"On one side," Vancouver wrote, "it presented the appearance of a sumptuous edifice fallen into decay; the columns which looked as if they had been raised with much labour and industry, were of great magnitude."

Vancouver's Pinnacles, as a cartographer at the time named them, became Pinnacles National Monument in 1908 and are now part of the national park system. The pinnacles lie thirty miles south of Hollister along the fairly straight rift-line route of State Highway 25. I have driven to the pinnacles from the Bay Area over the years when I wanted a quick hit of the Southwest, their place of origin having been the Mojave Desert.

A graduate student at the University of California at Santa Cruz, Vincent Matthews III, matched the pinnacles with the Neenach formation in 1973. Subsequent authorities have confirmed it, calling Matthews's work "one of the best documented ties across the fault." I used Matthews's *Pinnacles Geological Trail* guide when I recently hiked to the 2,700-foot crest of the rocks.

At 9 A.M. it was still cool when I set off on the Condor Gulch Trail into the heart of the pinnacles. I was ascending the east side, whereas Vancouver and his party had viewed the formation from the west. The trail began among the green grasses of early spring; Vancouver had been put off by the lack of verdure in November.

The dark red monoliths resembled the palisades of an abandoned fort or tombstones in a cemetery for giants that had been knocked askew by some tremendous force, like an earthquake. Purple lupine and scarlet Indian paintbrush grew along the trail, much like the remnants of a formal garden. I heard the downward trill of a canyon wren, a sound I associate with the Grand Canyon.

I passed through the phalanx of stone that guarded the heights, a wondrous gift from the interior of the earth via the fault line. Besides hikers like myself, rock climbers took advantage of these spaces.

I reached an overlook and had an orange and some water. The day was now hot. Matthews wrote of this vantage point:

> From here you can get an excellent view of the surrounding country-side. Looking northward, you see a large hill in the near distance which appears to have been cut in half. At the base of the hill is a line of trees. These clumps of trees appear where springs occur along the Chalone Creek Fault. The Chalone Creek Fault was part of the San Andreas Fault until a large chunk of sandstone jammed the San Andreas and caused it to move five miles east, to the approximate location of present day Highway 25.

From the heights I looked down upon the narrow band of trees and the much greater incision of the modern San Andreas Fault a few miles to the east. I made my way along the precipitous High Peaks Trail and descended to the Bear Gulch Visitor Center, where my five-mile-loop hike into these "most extraordinary mountains" had begun.

It is not possible to follow the San Andreas Fault all the way from Pinnacles National Monument to Parkfield. There was no reason to push roads into the less than sparsely settled canyons and ridges of the interior Coast Range. Peach Tree Road goes part of the way before a detour west to the 101 free-way is necessary. An alternative, whereby the fault line is abandoned earlier yet regained sooner, is State Highway 198 and then the dirt road, best not driven in wet weather, called the Parkfield Grade.

Regardless of the approach, the name fits the place. The majestic oaks form a green oasis of shade within the vastness of the exposed grazing lands. Parkfield is halfway between the San Joaquin Valley and the coast and fifteen miles north of the main east-west route, State Highway 46. You either live there or have a specific reason for visiting, as did the hordes of scientists and media personnel who periodically descended on the small hamlet hoping for an earthquake that, much like an ill-mannered guest, never did arrive on time.

Parkfield was the site of this country's first official earthquake-prediction experiment. It is where millions of dollars worth of monitoring equipment recorded little besides the centuries-old phenomenon of fault line creep and the constant chirp of crickets. It is the place where simple solutions collided with unfathomable complexities. At Parkfield, the people with advanced degrees thought they had discerned the faint terrestrial beat of a regular rhythm; and, accordingly, they spread their nets to catch the wily beast within the earth.

The first known earthquake struck in 1857. It may have been a smaller quake or a series of quakes in the Parkfield area that, as conventional explosives are used to trigger a nuclear detonation, set off the gigantic Fort Tejon temblor.

By 1881 there was a small settlement named Imusdale a short distance from what would become known as Parkfield. On February 2 there was a shock and chimneys fell to the ground, large cracks appeared in the road, and sulfur springs bubbled to the surface. As with the 1857 temblor, there were a number of foreshocks that would come to be regarded as reliable precursors to Parkfield quakes.

A story is told about this time. I heard it from Duane Hamann, the village schoolteacher who works part-time for the USGS. He said:

> A fellow by the name of Imus came out here with his brother and they lived down the way about five miles. I guess they had a scrap of some kind. So Imus moved down here by the bridge, right on the San Andreas Fault. An earthquake came along, and it shook down his chimney. He went outside and said, "Lord, if you are going to do this to me, put my chimney back up."
>
> Well, Parkfield has this twenty-minute warning on a lot of quakes—in other words, the foreshocks. And so he went about his business. Twenty minutes later a worse one came along, and he went right back outside and said, "Never mind Lord, I'll put it up myself."

The next earthquake was in 1901. A resident wrote his family that "poor old Earth trembled and groaned like some person in great agony, and at intervals ever since has rumbled and shook. All the chimneys in town were shaken down and the ground is seamed for miles, they tell me. Half the people around seem half scared to death."

Harold Fairbanks traveled through the area a few years later collecting information for Lawson's report on the San Francisco quake. He noted: "The region about Parkfield, in the upper Cholame Valley, has been subjected to more frequent and violent disturbances than almost any other portion of the entire Rift."

Twenty-one years passed. There was another moderate shock in 1922, followed by yet one more a dozen years later in 1934. The two quakes were detected by seismographs in Berkeley and the Netherlands. A comparison of the seismograms revealed a remarkable similarity.

For local residents, however, each quake was a different experience.

Around 3 A.M. on March 10, 1922, a young cowboy by the name of Buck Kester was sleeping in a small room attached to a house on the fault line. He heard the children in the family scream. The horses stampeded from the barn. The cowboy later remarked, "I needed spurs to ride the bed."

Kester, a veteran of four quakes, lived to the age of ninety-eight. He regularly supplied the media with local color. A friend told the story about the time scientists from the USGS approached Kester for permission to locate monitoring equipment on his ranch:

> They were talking in Buck's yard when a rattlesnake slithered toward them. The USGS people started looking for sticks or other weapons, but Buck calmly continued talking as he lifted one foot and stamped down with a boot heel squarely on the rattlesnake's head, killing it. He would later tell news reporters covering the experiment, "Earthquakes are just like rattlesnakes. I'm not afraid of them, but I can get along without them."

On June 7, 1934, there was the customary foreshock; seventeen minutes later the main quake hit. Nine-year-old Donalee Thomason was waiting to go on stage for the end-of-year school program. "I remember being thrown back and forth against the walls of the narrow runway behind the stage," she said. "It seemed the hall was turning upside down there in the darkness for a few seconds." There was running and screaming and a few children fell down, but in the best tradition of Broadway, the show went on—albeit rather shakily, as the aftershocks continued throughout the performance.

Elsewhere, chimneys again fell and the earth cracked along the fault line. A concrete-block house collapsed. Miners felt a strong shock six hundred feet below ground in Stone Canyon, just north of Parkfield.

The June 26, 1966, quake struck seventeen minutes after the first

noticeable foreshock and thirty-two years after the last temblor. Donalee
Thomason, now married and in a home of her own, recalled: "The very first
thing my ears recorded was similar to a great drawing in of a breath, or a suction sound may be a better description. Then a blast of hot air hit my back
as the shock wave rushed through."

By accident, the USGS was prepared for this quake and got excellent
readings. Seismographs had been deployed in the area to monitor an underground nuclear explosion in Alaska. For a shock in the low-to-moderate
range, as all Parkfield quakes were, "surprisingly intense effects" were noted
in 1966—effects that could be expected from quakes with one hundred
times the amount of energy.

There was a forewarning—a precursor, if you will—of what would occur
later in the century. Buried in the USGS report on the 1966 quake was the
following notation:

> The association [between released energy and physical effects]
> certainly indicates that the potential destructiveness of earthquakes
> of moderate magnitude should be reappraised, especially for seis-
> mically sensitive structures in alluviated areas near faults of the San
> Andreas type.

In 1966, headstones fell in the Parkfield cemetery, oak trees toppled in
fields, cars were picked up and deposited inches from where they had been
parked without a visible trace of movement, the usual chimneys fell, springs
dried up, water sloshed in cracked swimming pools, and floors and walls
pulled apart. The damage was minor because of the narrow hazard zone, the
resilience of wooden buildings, and the few structures.

The quake stirred up a great deal of excitement in the developing California seismic community. Except for the distant Alaska temblor of 1964, there
had been little to study. The next day scientists from the Menlo Park office
of the USGS and Caltech met in Parkfield. Robert E. Wallace of the USGS
gave the following account of the meeting:

> We found Clarence Allen [of Caltech] lying in the shade of a big live
> oak tree after an all-night stint of driving and field study, using head-
> lights and flashlights.

"Did you see the offset in the white line of Highway 46?" he asked.

"Yes, we did," I said, taking out my sketch of the offset. "It was about 5 centimeters," I reported, or nearly 2 inches.

"It was?" Clarence replied incredulously. "I measured only about 1 inch."

We soon realized that we had encountered a new, previously unreported phenomenon, now known as post-earthquake creep or slip. That was very exciting.

In the 1980s the USGS established the Parkfield earthquake-prediction experiment. Time passed, the predicted earthquake failed to appear, and the excitement evaporated. Duane Hamann spent less time now taking readings for the USGS. But he still conducted earthquake drills in the one-room Parkfield schoolhouse in the San Andreas Fault zone.

On a recent day, school began with an enthusiastic Pledge of Allegiance by a dozen students, whose grade levels ranged from kindergarten to sixth grade. The tall, blue-jean-clad Hamann moved from student to student giving each a great deal of individual attention. His comments were sprinkled with a lot of "I'll be darneds" and "you betchas." The students called him "mister."

Hamann was in his twenty-eighth year of teaching in Parkfield. He can cite former students who are doctors and lawyers. One of his former students was scheduled to bring her high school science class to Parkfield in a few days for a talk by Hamann and a tour of the monitoring sites. There was a problem, however, with some students not finishing high school.

I noticed the false ceiling that hid the heavy light fixtures in the room and thought of the near misses elsewhere because schools were not in session when strong quakes hit and such weighty objects came crashing down. Hamann was not pleased with the upgrading the school, built in 1951, had undergone a few years earlier. He termed it more a remodeling than a retrofitting. On his own initiative, Hamann had put clear plastic film over windows and bolted down bookcases and electronic equipment.

Because he lives in Parkfield, yet believes in the experiment, Hamann is of a divided mind about the desirability of an earthquake.

We stood outside during recess. Hamann pointed to the nearby fault and said: "I put in a little extra time because I have thirteen years of gathering data. I don't want the earthquake to sneak up and catch us unaware. I'm still hoping to see something on the instruments before the quake."

And so were a lot of other people.

THE SEISMIC SCIENCES IN THE TWENTIETH CENTURY

Earthquake prediction was referred to as the holy grail of seismology. Prediction was a chimera that came and went. Failure left the science in a tenuous position at the end of the century.

The twentieth century did not begin auspiciously for earthquake prediction. Gilbert's remarkably prescient address to the geographers in 1909 was made available to a wider scientific audience. The lengthy paper by the veteran of the 1906 earthquake was reprinted in the journal *Science*.

The theme of Gilbert's speech was earthquake forecasting. He outlined the necessary factors: "definiteness as to time and place." As to place, there were no real problems; faults, such as the San Andreas, were known, he said.

Time needed to be specific. To say that an earthquake was due over several days or months or even within a given year was not helpful. The problem, Gilbert said, was that "expectation would be tense, and the cost in anticipatory terror would be great." He foresaw no "precision attainable along lines of achievement now seen to be open." Gilbert said the successful determination of timing lay in "the indefinite future."

The difficulty was the lack of reliable "prelude phenomena," which came to be known as precursors. He discussed foreshocks and cautioned: "But unfortunately there are exceptions, and the character of the exceptions is not reassuring." The same pertained to earthquake weather, sounds, and "the peculiar behavior of animals."

Rather than prediction, the solution to dealing with earthquakes, Gilbert thought, was what would become known as earthquake engineering:

> If, on the other hand, the places of peril are definitely known, even though the dates are indefinite, wise construction will take all necessary precautions, and the earthquake-proof house not only will insure itself but will practically insure its inmates.

Gilbert was, however, mistaken on one count. Earthquake-proof structures do not exist. With the advent of earthquake engineering and the incorporation of its findings into modern building codes, there was a greater, not an absolute, degree of safety.

It took a while for earthquake engineering to become recognized as part of the engineering profession. It was a fringe field that was dependent upon a tangible response to a widely varying force whose parameters were not known. The data for the response and force have always been incomplete.

The first indications that earthquakes were not a natural disaster—that the human element of who had built what and where was a greater factor— came with the Fort Tejon earthquake. Although primarily a central and southern California affair, it was also felt far to the north.

The *San Francisco Bulletin* of January 9, 1857, commenting on the fore-shocks of the previous night, succinctly expressed that city's structural quandary: "What between fire and earthquake, it is hard to say whether wooden or brick houses are safest to live in." Wooden structures bent with the shock and were more resilient, but they were flammable. Brick buildings were rigid, but they were more fireproof. The shocks, the newspaper noted, "seemed to be much more severe in the lower than in the upper part of the city." The lower sections were on unstable ground.

After the 1868 Hayward quake, many engineers in the Bay Area used bond iron to reinforce masonry buildings. Charles Derleth of the University of California, who investigated the engineering aspects of the 1906 temblor, offered this opinion the following year: "An attempt to calculate earthquake stress is futile. Such calculations can lead to no practical conclusions of value."

Actually, there was little seismic information at that time on which to base calculations. Accelerographs were not developed until 1932.* A partial accelerograph reading was obtained from the 1933 Long Beach quake. But a useful reading was not recorded until the 1940 Imperial Valley temblor. Higher readings came with each succeeding urban earthquake. The engineers were surprised each time.

Earthquake engineering was a retrospective profession; the seismic elements of building codes were adjusted for readings from the last event. The

---

* Accelerographs are modified seismographs. They measure the sudden jolt of ground motion. Buildings need to withstand this shock. Seismographs measure the motion caused by seismic waves.

codes were enacted to prevent the collapse of structures and loss of life in a major quake. Yet some structures failed in moderate earthquakes, and there were deaths.

The building codes offered no magic formulas. The noted earthquake engineer Henry J. Degenkolb once said of the seismic requirements:

> Engineering is not a nice mathematical certainty. Engineering in earthquakes is even worse. There's a lot of judgment that goes in it, and I've often said, some of the main things in earthquake resistant design are not even mentioned in our codes because we don't know how to put them in the code.

Following the 1989 Loma Prieta quake, a governor's board of inquiry, headed by George W. Housner of Caltech, the dean of earthquake engineering, concluded:

> Two basic problems have been that the level of research effort on earthquake engineering has not been commensurate with the size of the problem and that the information provided by research lagged behind the need for it; the lag time for bridges [heavily damaged in the 1989 quake] was approximately 20 years. As a consequence, California is more vulnerable to earthquakes than it should be.

With each successive quake there were engineering successes; but there were also some remarkable failures, not only in new construction but also in those structures that had been retrofitted. A state report on the 1994 Northridge quake noted: "The shaking was strong enough over a large area to push structures beyond their elastic limits, exposing design and construction errors as well as obvious weaknesses in the code." This was not a major quake, and Los Angeles was the national leader in specifying engineering criteria for earthquakes.

Who was to blame? The engineers felt that the seismologists had the advantage in the battle for funds. The seismologists thought the engineers were too slow to adapt to new realities, such as higher ground-motion readings.

In the 1995 presidential address to the Seismological Society of America, Thomas H. Heaton of Caltech, who held a dual appointment in the seismological and engineering departments, said there was a third component to consider:

> Practicing engineers are paid by building owners, and it is difficult
> for them to publicly discuss their concerns about these buildings. Seis-
> mologists don't say much about the situation because they know so
> little about buildings. Although there are many interests to protect, who
> protects the interests of the occupants?

The answer was no one. There was no Ralph Nader or Common Cause or Sierra Club of earthquakes. It was mostly an intermittent regional problem, and the West Coast had other, seemingly more pressing, concerns.

The earthquake establishment, certainly minuscule when compared to other special-interest groups and internally fractious, relied on self-policing measures. For a group whose main concern should have been the practicalities of public safety, the profession was quite insular and overloaded with academically oriented practitioners.

For earth scientists, the years between the San Francisco earthquake and the 1960s were a time of theoretical stasis. It took an unusually long time for geologists and seismologists to accept a new theory that would give them a predictive tool.

In 1912, Alfred L. Wegener, who was outside the geological mainstream, proposed that the continents had drifted apart, an idea that dated from the sixteenth century. It took another sixty years for continental drift and its sibling, plate tectonics, to be accepted as gospel.

To Gilbert and others, earthquakes were caused by "the sudden breaking or slipping of rocks previously in a condition of shearing strain." Fractures occurred, said Gilbert, when internal stress exceeded the strength of the rocks. The result was a fault. The surface of the earth was conceived as being stationary.

Change was vehemently resisted by the more staid elements of the discipline, mostly located in the United States. It did not help that there were two world wars; and Wegener, a German, fought in one of them. His degree was in astronomy and his interests lay in the fields of meteorology, paleo-climatology, and polar exploration.

The Americans should have paid closer attention to Wegener. After all, it was Gilbert, the nonseismologist, who had advocated "the principle of scientific trespass." The narrow specialist, said Gilbert, missed the advantages of "cross-fertilization." Looking back at the controversy from the perspective of the 1970s, William Glen wrote in his book *Continental Drift and Plate Tectonics* that "the story of continental drift bears out the philosopher's charge of narrow scholasticism and close-mindedness against the scientific community."

There were, however, a few minor advances during the intervening years.

A gentlemen's agreement between seismologists at the University of California at Berkeley and Caltech in Pasadena (Perry Byerly for the former and Beno Gutenberg for the latter) was reached in the 1930s to divide responsibilities for earthquake research at the Tehachapi Mountains. Each university monitored its portion of the large state. When the USGS arrived on the seismic scene in the 1960s, the federal agency adapted to existing geoacademic politics and divided its activities and funding accordingly.

There was another development. While it was generally known that there was a single long fault, northerners held that there was little or no offset on the San Andreas. Nicholas L. Taliaferro, a Berkeley geologist who, along with his students, was the leading advocate of the little or no displacement theory, wrote in 1938: "The total horizontal movement is small, certainly less than a mile." Southern Californians chipped away incrementally at this opinion to where there is now agreement on offsets of around two hundred miles.

In 1946, the Jesuit seismologist J. B. Macelwane gave a nationwide radio address that was later published as the lead article in the *Bulletin of the Seismological Society of America*. The distinguished seismologist asked why there was no earthquake-forecasting service that was the equivalent of the Weather Bureau. He answered in a Gilbert-like vein:

> The problem of earthquake forecasting has been under intensive investigation in California and elsewhere for some forty years; and we seem to be no nearer a solution of the problem than we were in the beginning. In fact, the outlook is much less hopeful.

This prevailing pessimistic view was echoed by Charles Richter, who wrote in his widely used textbook, *Elementary Seismology:* "The student should be aware of the little significance attempts at prediction have had in relation to the actual development of our knowledge and understanding of earthquakes."

After World War II, new technologies became available and there was a new crop of scientists ready to employ them. Science was not immune to what was going on in the wider world; and the sixties was an explosive time, in many ways. Cracks began to develop in the theoretical dogma.

Oceanographers led the way, and the new findings were assimilated in mid-decade by a Canadian geophysicist, J. Tuzo Wilson. The new data showed that the earth's crust moved and that earthquakes on the San Andreas Fault were not a matter of simple slippage, as Gilbert and others thought. The San Andreas was promoted to the status of a transform fault, meaning it was the boundary between two of the seven major plates that composed the surface of the earth.

With an enthusiasm that matched the intensity of their earlier intransigence, American scientists embraced continental drift in the form of plate tectonics. Earth science was revitalized; it was an exciting time. Here was the theory for the movement at plate boundaries that explained earthquakes, although intraplate quakes remained a problem.

The emphasis was now placed on movement, and neocatastrophism (the old catastrophism having relied on God's will) challenged Lyell's gradualism. One of its chief advocates, the geologist Derek Ager, wrote in his book, *The New Catastrophism: The Importance of the Rare Event in Geological History:* "Nevertheless, it is obvious to me that the whole history of the earth is one of short, sudden happenings with nothing much in particular in between."

By the end of the century, orthodoxy had once again settled over the field. The geologists and seismologists I talked to could not imagine any theory replacing plate tectonics, despite the history of changing concepts dating back to Aristotle's time.

In his landmark book about scientific revolutions, Thomas S. Kuhn wrote: "No part of the aim of normal science is to call forth new sorts of phenomena; indeed those that will not fit the box are often not seen at all. Nor do scientists normally aim to invent new theories, and they are often intolerant of those invented by others."

Kuhn, a philosopher, was teaching on the Hayward Fault at Berkeley when he wrote *The Structure of Scientific Revolutions* in 1962. An enlarged edition was published in 1970. Yet there is barely a mention of geology and no discussion of plate tectonics, the perfect example of a revolutionary paradigm shift.

Geology was not regarded as a really serious science. Too oriented toward practicalities, was the snotty objection. The discipline has struggled self-consciously against such prejudices, which is perhaps why the more impressive term *geophysics* was invented and is now in vogue.

THE BIRTH OF THE MODERN PREDICTION EFFORT

With a theory in hand, all that was needed was an actual earthquake. The Alaska quake of 1964, the second most powerful to be captured on seismographs, was the event that jump-started modern earthquake prediction.* Robert Wallace of the USGS, a longtime observer of the political and scientific wars from inside the earthquake establishment, said the Alaska quake, centered in Anchorage, "started the modern era of earthquake awareness in the United States." The attempt to predict earthquakes became the tail that wagged the seismological dog.

The seismological community was ready to take off in a new direction. It had received a large infusion of funds from the Department of Defense in the 1950s and early 1960s. With a nuclear test ban looming and the Russians opposed to on-the-spot verification, millions of dollars had been poured into a worldwide network of 125 seismograph stations that could detect the difference between an earthquake and a nuclear blast.

When the job was done, the "money faucet," as Wallace referred to it, was turned off. The Denver USGS group that had been working on detection moved to Menlo Park in the mid-1960s and began to study earthquakes. Academic scientists who were also dependent on the military for funding were left "high and dry," according to Cinna Lomnitz, the author of *Fundamentals of Earthquake Prediction*.

* The 1960 earthquake in Chile is generally considered the most powerful, with three times the energy equivalent of the Alaska quake.

Lomnitz, one of Richter's Caltech students, was based at the Institute of Geophysics at Mexico City's Universidad Nacional Autónoma de México. Knowledgeable about the science yet outside the inner circles, his views on prediction have a certain perspective that others lack. Lomnitz characterized prediction as "the very soul of seismology"—meaning that it draws upon all of that science's basic elements—and thought it was mostly politically driven. He wrote in 1994:

> Minerva, the goddess of wisdom and science, sprang fully armed from the recesses of Jupiter's brain. Similarly, earthquake prediction was born of the deep pockets of government science in the 1960s and 1970s. The bounty has dried up, but the birthmark remains.

The Russians had reported some successes with prediction; and the Japanese, ever the leaders in seismic research, launched an ambitious program in 1965. (Their leadership had been acknowledged as early as 1909 by Gilbert. They led for good reason; earthquakes were an integral part of the nation's culture. Two-thirds of the capital city of Tokyo was destroyed and 140,000 people died there in a 1923 earthquake and fire. I wonder what this country's response to seismic events would have been if Washington, D.C., instead of San Francisco, California, had been nearly destroyed in 1906.)

A working group of American seismologists, headed by Frank Press of Caltech, cast about for a national program following the Alaska quake. They came up with a ten-year, $242 million proposal. Unfortunately, the group noted in an appendix to their report that the goals could also be met with "existing knowledge," a strategic blunder and a hubristic boast.

Although they were involved with the deepest history of the earth, a historical perspective of their own discipline was not a forte of the earth scientists. For instance, there is no history of earthquakes, the science of earthquakes, or earthquake engineering. They tended to focus on minuscule aspects of earthquakes or, at best, a single event.

An exception was Wallace, whose career spanned the last half of the century. Wallace was unusual in other ways; he associated with earthquake engineers and held both high-ranking management and research positions within the USGS. Wallace noted in an oral history:

> Moreover, from the moment that the Press report championed prediction, that idea seemed to dominate. Even the sociologists hung their program proposals on prediction. The seismological community, strong proponents of prediction, dominated within the earth sciences in terms of budget distributions. An important factor, I believe, was that Congress, at least some members, were turned on by the idea of earthquake prediction—a high-tech approach in a high-tech era. In my opinion, the whole earthquake program might not have gotten off the ground without the excitement generated by the idea of prediction.

Optimism about earthquake prediction was rampant within the scientific community in the mid-1970s. Such feelings were certainly genuine, quite expedient, and a complete reversal of what had prevailed earlier in the century. However, the opinions voiced at the highest levels of the profession lacked the hard data and hard-nosed peer reviews that scientists seemingly insisted upon before coming to even minor, let alone major, conclusions. This absence of process was a travesty of science.

The authors of a study of research on natural hazards, funded by the National Science Foundation (NSF) and published by MIT Press in 1975, stated that "specific forecasts of damaging earthquakes" might be possible the next year. They were not sure whether such forecasts would be a blessing or a curse, given all the possible societal consequences.

A National Academy of Sciences (NAS) panel headed by Clarence Allen of Caltech issued a report in 1976. Citing the "great excitement" prevalent "among seismologists over major achievements in our efforts to predict earthquakes," the report mixed optimism with caution and noted the lack of requisite knowledge:

> A scientific prediction will probably be made within the next five years for an earthquake of magnitude 5 or greater in California. With appropriate commitment, the routine announcement of reliable predictions may be possible within 10 years in well instrumented areas, although large earthquakes may present a particularly difficult problem. The apparent public impression that routine prediction of

earthquakes is imminent is not warranted by the present level of scientific understanding.

A 1978 NAS report on the socioeconomic effects of prediction repeated the one-decade estimate and added: "The idea that it may become technically possible to predict earthquakes has been recognized by two reports from the National Academy of Sciences as well as by numerous scientific and popular publications." Two years earlier the chairman of this panel, Ralph H. Turner, a sociologist at the University of California at Los Angeles, had written that predictions "specifying the place, time and magnitude of future quakes" were imminent.

In the midst of all this hoopla, Congress passed the National Earthquake Hazards Reduction Act in 1977. It was the first and only national earthquake initiative. The goals of the act were spelled out in the National Earthquake Hazards Reduction Program (NEHRP) sent to Congress by the Carter administration the following year. The principal goal was to reduce losses from earthquakes.

Besides the USGS, which was put in charge of prediction and received the bulk of the funds, three other federal agencies were involved in the program: the National Science Foundation, the National Institute of Standards and Technology, and the Federal Emergency Management Agency. The latter was responsible for overall coordination.

Earthquake research undertaken by this umbrella group was one of the smaller federal programs. By the mid-1980s, the funding amounted to $60 million a year, of which $17 million was earmarked specifically for prediction. It rose to about $100 million in the next decade, with half going to the USGS. Other agencies, such as the Nuclear Regulatory Commission and the Federal Highway Administration, had separate seismic programs that fit their special needs. Their joint funding level was more than three times that of the NEHRP agencies.

There were problems with the program.

Henry Degenkolb, the earthquake engineer, said in the mid-1980s: "The big concern of NSF, and a concern of a lot of us, is that while we are doing a lot of research—and while a lot of it is chaff, I'll grant—it's also very difficult to get significant research results put into practice."

Looking back from the vantage point of 1996, Robert Wallace said that "the politics of personalities and power" were the major factors. Good science was not mentioned. He described the internal struggles of the previous two decades:

> There was a battle between the engineers and earth scientists for dominance in the program, and within the earth sciences there was a battle between seismological strategies and geologic strategies.... But there never has been a way to divide up money so as to please everyone, nor did the contest end with the passage of the National Earthquake Hazard Reduction [Act].

Wallace, now a USGS scientist emeritus, portrayed the policy and funding debates within the scientific community as an unruly process. Personalities, rather than science, were at the center of policy and funding discussions. He cited "the strong personalities," "clashes," "harsh words," "strong feelings," and "strong animosities" that were involved. At the core of the various controversies was how the money would be divided. "The rationale that 'only seismology equals earthquakes' seemed to rule the day in the USGS," Wallace noted.

There is a document that provides an unusual look at how the profession viewed itself on the eve of the present decade. Seven months before the 1989 Loma Prieta earthquake an anniversary dinner was held in the Faculty Club on the University of California's Berkeley campus—the same place where Gilbert was awakened by the 1906 earthquake. A distinguished group of seismologists, geologists, and earthquake engineers gathered to look back on twenty years of accomplishments in California.

The celebrants included Clarence Allen of Caltech; Bruce Bolt, director of the Berkeley seismological station; Henry Lagorio of the university's School of Architecture; Lloyd Cluff, chairman of the California Seismic Safety Commission; Thomas L. Tobin, the commission's knowledgeable executive secretary; Robert Olson, head of a Sacramento consulting firm; and Degenkolb and Wallace.

The unusually frank discussion, which was recorded and later published in an obscure document, got around to the heading "Evidence of Professional Rivalry." The title was assigned to that portion of the text by its editor, Stanley Scott of the university's Institute of Governmental Studies. Scott was also a member of the state commission and was present at the dinner.

The discussion follows:

Tobin: I worry about the field of seismology because I see there are so many who seem to be at one another's throats.

Allen: That's the way we look at the engineers.

Tobin: It may be, but that puzzles me.

Allen: Yes, I don't understand it.

Tobin: I see so much of that. It seems there is so much backbiting among the competition, that I don't see any of the younger people really rising to the surface. I see all these people, but I don't see any leaders coming out of it.

Allen: Maybe I don't see the leaders, but I am surprised at the backbiting thing. I really am.

Cluff: I think what we're seeing from [the commission's] research committee activities is a group of people who are not seismologists. They are geophysicists who see an opportunity to get into some funding, and they're a bit confused.

Olson: There's nothing like the availability or scarcity of money to make cleavages real quick.

Cluff: I think that's part of it.

The topic next switched to "Disagreement in Science":

Wallace: There is disagreement among the earth science disciplines because the science of earthquakes is emerging rapidly, and we're bound to disagree with each other. That's a healthy and necessary part of the scientific process.

Cluff: Yes, but isn't the public confused?

Olson: No, I don't think they seem confused.

Lagorio: You give one magnitude here and another magnitude there.

Tobin: I see different agendas. One group says that the most important topics are blind thrust faults and another group disagrees. Whereas every year the structural engineers have to put together the Blue Book. In the process they might disagree with each other, but they agree on something at the end.

Wallace: We are much more fluid in the earth sciences and appreciate the fact that ideas do evolve.

Earthquake prediction certainly did evolve. It had its downs, its ups, and currently it is out of favor. However, prediction is still very much on the public's mind. When people asked me what I was working on, and I said a book about earthquakes, invariably the response was: "Well, when is the next one?"

The best way to evaluate prediction—and thus, in large part, the performance of the earthquake sciences in the last quarter century—is to examine two case studies: the Chinese experience and the Parkfield experiment. A discussion of earthquake phenomena is part of the predictive mix.

THE CHINESE EXPERIENCE

Each country's basic approach toward earthquake prediction differed in accordance with its tectonic history, geography, and culture.

The United States has two hundred years of earthquake history; China has three thousand years. Of the fifteen thousand citations concerning seismic events in the written records between the years 1177 B.C. and A.D. 1956, a Chinese seismologist commented, "Such long and rich records could never be found in other countries." In the United States, the casualties, mostly confined to one state, numbered in the low thousands; in China, they were in the low millions—thirteen million by one count—and were spread throughout the country. As a result, the Chinese have a national earthquake culture.

The Chinese have pursued prediction more assiduously, but neither nation has mastered it. There has been one formal prediction in the United States, and it failed. The Chinese claim a 25 percent success rate for numerous predictions. An official of the State Seismological Bureau said in 1996 that "generally speaking, the failures by far outnumber the successes in prediction efforts of the past thirty years."

Geography is a factor in seismic prediction. Quakes can occur across the breadth of China and Tibet and professionals are spread thinly over the most likely areas. Thus, amateurs are needed. Where communications are poor, a more locally oriented decision-making process makes sense. In this country, most of the quakes occur in California, and the professionals are clustered in that state. The funding and directives come from Washington,

D.C. The one formal prediction was endorsed first at the federal level and then seconded by the state.

Greater wealth allows a society the luxury of engineering for earthquakes, building more resistant structures, and reliance on technological fixes. Thus, casualties have been far less in America, given the same magnitude and number of shocks. There are more people in China and less wealth and modern technology. In China, seismology is a people-driven science focusing on preventive measures. In the United States, the science is more instrument-driven and aimed principally at research.

Folklore is more deeply embedded in the older culture. A U.S. delegation investigating the 1975 Haicheng earthquake stated in its report:

> But the emphasis on animal behavior, water well anomalies, meteorological changes, earth lights, and earth sounds is based on folk tradition and on written accounts. This approach contrasts with the American experience, in which animal behavior anomalies preceding earthquakes have often been reported popularly but are generally dismissed without serious investigation by seismologists.

There is a traditional belief in China that earthquakes are a harbinger of governmental upheaval; thus, the rulers have a real stake in preventive measures that is absent in this country. Lomnitz cited the following example:

> Chang Heng (78–139), the inventor of the first seismograph, notified the Emperor of the occurrence of a large distant earthquake three days before the news reached the Court, thus enabling the sovereign to claim foreknowledge. It might otherwise have cost him his throne. Understandably, the "struggle against superstition" carried out by earthquake brigades between 1966 and 1976 was largely directed against such a dangerous belief.

The first earthquake recorder, constructed in A.D. 132, was an ingenious affair. Eight dragons with copper balls in their mouths were stationed at points on the compass. Eight frogs with their mouths open waited below. With the arrival of the seismic wave, a pendulum dislodged the corresponding metal ball, the noise alerted the observer, and the epicenter of the quake was determined by which ball had dropped. An earthquake more than four hundred miles distant was supposedly located in this manner.

From such an auspicious start, the Chinese came to lag behind the more technically advanced, seismically prone nations in the first half of the twentieth century. Nine major quakes from 1966 to 1976 thrust the People's Republic of China into the modern seismic era.

Premier Chou En-lai traveled three times to the site of the first devastating quake in the short history of the People's Republic. He exhorted the scientists in 1966 to solve the problem and was impressed by the stories of precursory phenomena, such as animal behavior. China attacked earthquakes with its greatest resource—people organized in military phalanxes. "Trust the people" was the defining slogan of the Chinese Revolution.

Nine years later China electrified the seismic community with the first successful prediction of a major earthquake. "The story of Haicheng has played a role in buttressing bids for government support to earthquake prediction programs in other parts of the world," wrote Lomnitz, who thought the glowing accounts in the West were tinged with expediency.

Long familiarity with earthquakes aided the Chinese. Their history contained many accounts of abnormal phenomena preceding earthquakes. As a result, they monitored the weather, behavior of animals, earth sounds, earth lights, groundwater levels, and foreshocks. All were considered possible precursors.

Given this tradition of seismic awareness and the rapid escalation of all the harbingers, the successful prediction of the Haicheng earthquake was hardly surprising.

Haicheng was a city of ninety thousand inhabitants some 350 miles northeast of the capital of Beijing in Liaoning Province. It lay on the old Russian-built railroad line that ended at what was once known as Port Arthur at the end of the Liaodong Peninsula.

There had been a slight increase in seismicity since 1972, and there was a further surge in the last months of 1974. But there were few seismographs in the province, so the numerical record was incomplete. The entire peninsula south from the North Korean border was considered ripe for a moderate temblor, but no specific time or place was specified.

The winter weather at the latitude of North Korea was unusually hot. On the day of the quake the temperature shot up fifty degrees to register a highly unusual sixty degrees above zero. This was referred to in later Chinese scientific literature as a "heat storm," or "urban hot island effect." There was a strange layer of ground fog—described as "thick, patchy black, and white"—and bad smells.

Six hundred cases of strange animal behavior were reported in the forty-five days prior to the quake. Hundreds of hibernating snakes, tricked by the hot weather, left their burrows. Mice showed no fear of people and appeared in large numbers. Obedient dogs disobeyed their masters. The unusual behavior was noted in twenty species of domestic and wild animals. Most of the erratic conduct occurred within the two days prior to the quake, when the foreshocks escalated.

Shortly before the earthquake struck on February 4, 1975, an express train was approaching a station near Haicheng. In A *General History of Earthquake Studies in China*, published by Science Press in Beijing, there is the following account:

> Very bright purplish red flashes suddenly glittered in the dark sky before the train, the drivers immediately pulled down the emergency brake. Just as the train stopped the strong earthquake came, and the ground was shaking violently. Although the train was located in the [epicentral] area, it was saved and a very serious disaster was avoided.

There were 161 cases of rapid changes in groundwater levels. Water was muddy in some wells and the taste changed. The temperature of hot springs fluctuated. One well-known hot spring stopped discharging three times in the days before the quake. Wells turned artesian. Geysers of water broke through the ice.

The foreshocks were the most noticeable of the precursors and were the immediate impetus for the local authorities to act. During the first three days of February there were intermittent, small quakes. After 6 P.M. on February 3, there was a recognizable series of foreshocks at the rate of up to twenty an hour. After midnight there was another surge of activity, with more than one hundred small quakes an hour. Residents were awake throughout the night and in a state of panic. The earth tremors sounded like artillery fire.

At 10 A.M. the next morning, the provincial government issued an official alert. The seismic activity eased off around noon. At 7:36 P.M. the large quake struck. Even with the alert, 1,328 people died and 16,980 were injured. It was bitter cold that night. Perhaps so many died and were injured because they thought it was safe to return to their homes after the initial burst of activity subsided.

The residents did not flee Haicheng. Without waiting for an official alert, many had already built temporary earthquake huts. Such structures, made of any lightweight materials available, were the equivalent of the tent encampments that are erected in city parks after earthquakes in this country. Lomnitz, who saw the huts in Beijing following the Tangshan earthquake, said they were "built in front of the home according to age-old Chinese custom, whenever foreshocks are felt."

The U.S. seismological delegation concluded admiringly:

> The prediction of the Haicheng earthquake is an extraordinary achievement of the seismological and geophysical workers of the People's Republic of China, whose national program in earthquake research was less than 10 years old at the time.

Great success was followed by tragic failure, which was not as widely trumpeted by either the Chinese or American seismologists, for different reasons.

Eighteen months after the Haicheng earthquake, while the achievement of the Chinese was being hailed around the world, no foreshocks and no prediction preceded the Tangshan earthquake, which struck with nearly the same force as a magnitude 8 earthquake.

Tangshan was a thriving, highly polluted industrial city of one million residents just to the east of Beijing. China's seismic agency described the early-morning earthquake of July 28, 1976, thusly: "People were fast asleep when the earthquake struck. Lightning flashed across the sky and the earth rumbled ominously seconds before the ground began to shake."

It took about ten seconds to level most of the city. At the very least one-quarter of a million people died, a number that exceeded those both killed and injured in the nuclear blasts that leveled Hiroshima and Nagasaki.

Fortunately, there were few fires. One nearby dam almost collapsed. All that the embarrassed Chinese would initially acknowledge was that there had been a major earthquake, hardly a surprise since Western seismographs had detected it. Foreigners were not allowed to visit Tangshan.

What went wrong? The precursors were more subtle than those registered at Haicheng, and they varied. Timidity also ruled the day. The Tangshan area, which took in the capital, was densely populated; it was the most important economic, cultural, and political region of China. To shut down small Haicheng was one matter; Tangshan was another. The State Seismological Bureau noted twelve years later: "Most important of all, because of the political and economic significance of the area, any posting of a warning had to be carefully considered in light of its great social consequences."

The ancient belief that earthquakes foretold revolution almost came to pass in what was a remarkable demonstration of the psychological effects of a natural disaster and the political, social, and economic ramifications.*

Turmoil was abroad in the land in 1976. For ten years the country had been socially disrupted by the excesses of the Cultural Revolution and physically battered by upheavals in the earth. There were two purges in the party leadership; and Chairman Mao Tse-tung, Premier Chou En-lai, and Red Army leader Zhu De died that year.

The State Seismological Bureau said: "Psychologically, people had grown to believe in the invincibility of the scientists' ability to predict earthquakes." That faith evaporated. The Tangshan earthquake "elevated the latent fear of earthquakes to an irrational phobia among the people in China in the years that followed. This phobia had in places led to social disorder."

The remarkably clear and revelatory 1988 study by the government agency in charge of seismic concerns added: "At one time panic swept across the whole nation."

The report continued:

> Millions of people left their houses and moved to earthquake shelters set up on roadsides. All the materials used to build such shelters—wood, bamboo, mats and asphalt felt—were distributed free of charge by local governments. This not only caused great financial loss to the

---

* China was not alone in this regard. The weakening of Mexico's ruling party dates from its inability to deal with the 1985 Mexico City earthquake. Community organizers took over. A member of the Mexican Congress said, "It was a social temblor that affected everything."

government but often burglaries, fires and panic-purchase resulted, causing further chaos to society and normal social functions being disrupted.
At one point in time 700,000 Beijing residents stayed for days in tents, and
even foreign embassies reduced their staffs fearing an imminent disaster.

It took this colossal failure for the Chinese to realize the "diversity and
complexity" of the factors that went into a successful prediction effort.
Three trenches were dug across the fault and evidence was found for two
large prehistoric earthquakes with an average recurrence span of 7,500 years.
With some additional safeguards, Tangshan was rebuilt on the same site and
now the population is greater than before the quake.

EARTHQUAKE PHENOMENA

A high-level delegation of academics, government officials, and engineering consultants from the American earthquake community visited China
for three weeks in 1978. Although the delegation was not allowed into Tangshan, it was briefed on what had occurred there and on a later successful prediction effort in Sichuan Province. Explicit anomalies, such as a display
of earthquake lights, had preceded the series of three quakes in that
province.

The delegation's report on the Sichuan quake, published by the National
Academy of Sciences, contained one of the most detailed descriptions of
such lights:

> Earthquake lights, at least some of which were seen prior to
> the earthquake, were described in the forms of columns, fans, balls,
> and sheets. Chu Chieh-cho of the Provincial Seismological Bureau
> described personally seeing a fireball 75 km [45 miles] from the epicen-
> ter on the night of July 21 while in the company of three professional
> seismologists and a TV crew. The fireball originated at the ground sur-
> face about 100 m [300 feet] from where he stood. At first it was about
> 1 m [3 feet]. It then shot up to a height of 10 or 15 m [30 to 50 feet],
> whereupon the volume started shrinking to ping-pong ball size. After
> reaching the maximum height, the ball curved over in an arcuate trajec-
> tory and disappeared as it fell to the earth; resembling a meteor in

appearance as it moved. The light would dim, then brighten again. Small wisps of white smoke swirled around the light, a slight crackling sound was heard, and an odor of garlic or sulfur was detected. A radio compass and telluric currents were unaffected. A small funnel-shaped hole in the ground was found where, it was thought, gas had been blown out of the ground.

A member of the delegation, Robert Wallace, later commented in a journal article: "Considering the number of lights reported in the western world as UFO's, one might wonder what the reaction would be if the western populace were more fully interested in the possibility of observing sky lights and fireballs as phenomena premonitory to earthquakes."

The attitude that such phenomena were in the same category as UFOs was, and is, fairly commonplace among scientists in this country. But are they?

Nothing surprised me more in researching this book than the multitude of historical accounts of earthquake phenomena, about which I had previously known very little. I should state that I have never seen earthquake lights nor heard earthquake sounds. (I thought I heard distant thunder one clear day while writing this book. I ducked under my desk, but nothing happened.) Nor have I witnessed strange animal behavior or felt that the weather was unusual enough to indicate that an earthquake was in the offing. However, I have learned to trust disparate accounts of the same events that have common threads—at least enough to pass them on to others.

As the Chinese learned the hard way and the Americans would appreciate in time, none of the precursors occurred all of the time; but enough occurred part of the time to deserve serious study. I asked a respected seismologist why these types of phenomena had been studied only minimally in this country. That person angrily shot back: "What would you have me study?" Surely, that was hardly the question for me to answer. What happened to scientific curiosity? I wondered.

Another seismologist enlightened me. It was a matter of not enough money to spread around what were considered the backwaters of the discipline. As Helmut Tributsch, a European scholar, put it in his book, *When the*

Snakes Awake: "Any scientist interested in studying this problem would risk not only his reputation but also any chance of getting support."

There is a long history of such perceived events that dates back at least to Aristotle. Following the Lisbon and New Madrid quakes, Sir Charles Lyell summed up the Western world's knowledge of earthquake phenomena in 1830. *Principles of Geology* contains a list of precursors that "have recurred again and again at distant ages, and in all parts of the globe." Lyell wrote:

> Irregularities in the seasons precede or follow the shocks; sudden gusts of wind, interrupted by dead calms; violent rains, in countries or at seasons when such phenomena are unusual or unknown; a reddening of the sun's disk, and a haziness in the air, often continued for months; an evolution of electric matter, or of inflammable gas from the soil, with sulfurous and mephitic vapours; noises underground, like the running of carriages, or the discharge of artillery, or distant thunder; animals utter cries of distress, and evince extraordinary alarm, being more sensitive than men of the slightest movement; a sensation like sea-sickness, and a dizziness in the head, are experienced.

With the deployment of seismographs around the turn of the century, such phenomena came to be regarded as witchcraft. Edgar L. Larkin of southern California's Lowe Observatory traveled throughout the Bay Area after the 1906 earthquake and collected observations. Of Sonoma County, Larkin wrote, "The display of blue flames before the onslaught of the red ones, and their final yellow sequences, was very remarkable. The appearance of blue lights was over a wider area than first thought."

Larkin's descriptions of colored lights and other phenomena were published in the journal *Science* in July of 1906. A few weeks later Stanford's David Starr Jordan, a champion of the new objective science of geology, derided Larkin in the same journal. Jordan described Larkin as "professor of astronomy and geology in the University of the Sunday Supplement."

Referring to Larkin's descriptions of abnormal animal behavior, Jordan wrote:

> It was in San José, also, three days before the earthquake, so I am informed, that a cat was heard to utter three sounds sharp and high,

these followed by a hiss as of escaping steam. A dog was present and appeared also agitated. It was noticed that the dog's nose was cold, while the tail of the cat was rigidly erect.

The authoritative study of the state commission, published in 1908, cited eighty-one reports of earthquake noises. Also mentioned were lights and animal behavior. Cats fled and dogs barked. "The most common report regarding the behavior of dogs was their howling during the night preceding the earthquake." Horses whinnied and snorted in their stalls before the shock. It was milking time at the many ranches in the Bay Area. "Several instances were reported where cows stampeded before the shock was felt by the observer," said the commission's report.

In its assessment of earthquake lights, the National Academy of Sciences' committee on seismology leaned toward disbelief in 1933, but added a grudging caveat. The committee noted:

> Such effects are reported frequently, but the reports are never substantiated thoroughly. The possibility is strong that the apparent perception of electric flashes or luminosity, is subjective—but it is not impossible that such phenomena are actually produced.

The first quandary was what reports to trust. The second, assuming there was enough interest in the scientific community, was how to "capture" such activity, given the unpredictable and infrequent nature of earthquakes.

For instance, what could possibly have been in place and functioning in Sonoma County on the night of October 1, 1969, to document the multicolored lights that accompanied two moderate earthquakes. It was unlikely that any experiment would have been set up in advance, given Jordan's derisive comments concerning a similar occurrence in that same county in 1906.

Yet something did happen. It took the Santa Rosa newspaper one month to establish that there had been an episode of earthquake lights. On November 1, the *Press Democrat* reported in a page-one article:

> This is a story that has no ending—yet.
> It is about the peculiar and puzzling companion to the two hard earthquakes that struck Santa Rosa the night of Oct. 1—a series of brilliant flashes of light.

The basis is a flood of reports from witnesses in response to a *Press Democrat* request, reports that are strong evidence that the lights were more than just the flashing of shorted power lines.

With those reports came a number of startling "fireball" sightings before the quake hit, a doubly puzzling thing that some serious earthquake researchers have noted in other temblors.

But even the newspaper felt compelled to make light (as I just did) of the story that ran under the jiggly type of a headline that attempted to simulate the shaking of a quake. The comic-book-style drawing that accompanied the story showed the dark hills behind the city illuminated by what was labeled "FLASH!"

Twenty-five years ago a summation of what was known about earthquake lights, which is still considered one of the most complete, appeared in a ten-page article in the *Bulletin of the Seismological Society of America*. John S. Derr, who also couldn't resist a pun, termed such lights "the darkest area of seismology." He quoted an old Japanese haiku in his review of the existing literature on the subject:

> *The earth speaks softly*
> *To the mountain*
> *Which trembles*
> *And lights the sky*

The Japanese had collected the most material, including reports and photographs, of the lights. Derr's article was accompanied by one page of color and four of black-and-white photos. He concluded that "the existence of the phenomenon is considered well-established, although no completely satisfactory explanation has been advanced to date." The lights were probably due to some type of electromagnetic discharge, he said, such as that produced by rapidly compressed rocks. If that were true, Derr said, "it may be possible to develop monitoring methods for earthquake prediction."

To bring this account of one type of phenomenon full circle from the initial association of earth lights with UFOs, the air force's 1969 report, *Scientific Study of Unidentified Flying Objects*, contained a two-page discussion of "earthquake-associated sky luminescence."

Tracing accounts of such lights back to the Roman historian Tacitus, who described the Achaean earthquake of 373 B.C., the report concluded: "That large electrical potentials can be created by the slippage or shearing of rocks is not surprising. Of possible importance is the use of electrical measurements to provide some advance warning of an impending earthquake."

Earth sounds also had a long history.

Native Americans in California associated thunder with earthquakes in their folklore. It was Charles Davison, the British seismologist, who produced the definitive study of earthquake sounds. Davison consulted twenty thousand reports. He then categorized seismic sounds by types: most common was a sound like wheeled vehicles crossing a hard surface, second was distant thunder, third was a roaring or howling wind, sixth was the boom of distant cannons, and seventh was "an immense covey of partridges on the wing."

Audible sounds did not accompany every tremor. Davison determined that 43 percent of the earthquakes occurring in New England between 1638 and 1870 were accompanied by sounds. The noise produced by the Charleston, South Carolina, quake of 1886 was heard eight hundred miles from the epicenter. His study was published in 1938.

What is known about such sounds is slight, because investigation has been minimal. The noise, it is thought, is produced by the faster P wave, which arrives just before the stronger S wave that contains the shaking component. With the wavelength of the acoustic signal and the ground motion in phase, the earth acts like an enormous loudspeaker.

Not all such noises are within the range of human hearing. If the epicenter is nearby, then the two waves will arrive almost simultaneously, and the sound cannot be used as a warning device. (If I hear such a noise again on a clear day, I will not duck. But I will think "possible earthquake" and consider what to do, should the noise be followed by shaking. In that way I will not be overcome by the unexpected. I consider this my personal warning system.)

Earth sounds are the most frequently mentioned of all the phenomena by those persons who have experienced an earthquake. Of the 1994 Northridge shock, *Los Angeles Times* columnist Al Martinez wrote: "A deep-throated roar awakens me. It is something I had never heard before, a sound from the bowels of the earth, angry and threatening." My mother-in-law heard a similar noise in her San Fernando Valley condominium just before it was damaged and the occupants fled the four-story structure.

The concept of earthquake weather dates back at least to Aristotle, and was cited by Kant after the Lisbon quake. Lyell picked up the idea, and it spread into the current century through such means as word of mouth and popular literature. We regularly use the expression in California to describe hot, still weather. One of the many small quakes that occur each week may coincide with the observation.

I would like to believe in earthquake weather because it provides the perfect dramatic setting for an apocalyptic event. The common denominators in accounts are heat and oppressiveness. The smell of gas or sulfur compounds precedes or accompanies some quakes. A fog or slight mist might blanket the land, and sometimes the dust rises, lightning flashes, and heavy rains descend. Wow, what a scene!

While such weather remains the most nebulous and least explored of the precursors, there is no doubt about the psychological aftereffects of such catastrophic events on humans. Beyond nausea, there is lingering trauma, defined as "a semipermanent source of inner terror."

Following the nearby Loma Prieta earthquake, when the main Stanford library was severely damaged, university library workers were badly shaken and took advantage of counseling services. Seventy-five percent reported physical symptoms and over 90 percent had emotional difficulties, collectively defined as posttraumatic stress syndrome.

The young and elderly are most vulnerable to acute stress. David Alexander of the University of Massachusetts wrote in his comprehensive study, *Natural Disasters:* "The more sudden, unexpected or catastrophic the impact, and the less trained the victims, the more profound the syndrome." He added the intriguing thought, without any specific reference to California, that "high awareness of persistent or recurrent threats can lead to the development of 'disaster subcultures.' "

The contributions of sociologists and psychologists in this field have been slight. Bruce A. Bolt of the University of California at Berkeley wrote in his textbook, *Earthquakes:* "Despite the ample descriptions of the effects of earthquakes on buildings and the Earth's surface around the world throughout many centuries, there is surprisingly little information of the human reaction and social results."

Unusual animal behavior is the best-known of these precursors. After the San Fernando earthquake of 1971, two-thirds of southern Californians believed that animals could predict earthquakes.

The Japanese determined that animals, mainly fish and birds, were more sensitive to certain geophysical changes associated with earthquakes than humans. The variables that might affect them were the magnetic field, sounds (outside the range of human hearing), telluric currents (static or alternating currents flowing through the ground), ionization (radon emitted from the earth), the electrical field (such as whatever produces earthquake lights), and foreshocks (those too weak to be detected by humans without the aid of seismographs).

Following the Chinese success in forecasting the 1975 Haicheng temblor, America's interest in the less scientifically verifiable and tangible aspects of earthquake prediction peaked at a low level in the late 1970s. The USGS funded three modest studies of animal behavior that, interestingly enough, noted some correlations.

In the first study, three researchers from the University of California at Davis interviewed members of fifty households in Willits after a moderate quake in 1977. There were seventeen reports of unusual animal behavior before the quake in the northern California town. That "led us to conclude that some animals responded to physical precursors," they reported. However, hardly anyone in a small Montana town reported such behavior after a similar-sized quake in 1978. All lived much farther from the epicenter than the Californians.

In another experiment sponsored by the USGS, small rodents were housed outside in artificial burrows or inside in cages near the southern California desert community of Landers. A swarm of moderate-sized quakes struck in mid-March of 1979. The UCLA researchers reported: "To our obvious delight, significant, and in some cases dramatic, activity anomalies were recorded from most animals in both inside and outdoor facilities during the days immediately preceding and following the commencement of the Landers earthquake swarm." Further observation was recommended; but the experiment was terminated.

The most ambitious of the three studies was designed to mimic the Chinese experience. Seventeen hundred observers who lived in known seismic areas from Humboldt County to San Diego were recruited by Stanford

Research Institute International. They were told to call a toll-free telephone number whenever they noticed "unusual animal behavior whose cause was not immediately observable and obvious."

"A significant number" of such reports preceded seven earthquakes, but researchers said the results were inconclusive because they could not rule out other causes, such as "naturally occurring environmental or physiological events." Therefore, a congruence "of a large number of reports and an occasional earthquake may have been coincidental." Further study was recommended but never undertaken.

By 1985, when the Stanford researchers submitted their final report, the USGS had already embarked upon its own prediction experiment, which was supposedly grounded in firm scientific principles.

PARKFIELD AND THE POLITICS OF PREDICTION

Politics and bureaucratic infighting combined to produce the badly flawed Parkfield experiment at a time when the Chinese had lost faith in earthquake prediction. But the Americans thought they were on to something at Parkfield.

Shortly before the genesis of Parkfield there had been a blizzard of predictions in southern California, the most notable of which was the Palmdale Bulge. Parkfield followed the bulge, and both were rooted in history.

The Los Angeles region had a tradition of prophesiers of doom dating back to William Money (pronounced Mo-nay), who foretold in the third quarter of the nineteenth century the coming of comets, earthquakes, and the fiery destruction of San Francisco. Money drew a map of the world. Historian Carey McWilliams later wrote, "On this map San Francisco, a community that he detested, was shown poised on a portion of the earth that he predicted would soon collapse, precipitating the city into the fiery regions."

Such a prophecy was bound to be realized sooner or later, given the two Bay Area earthquakes in the 1860s and the fact that most of San Francisco had burned to the ground six times between 1849 and 1851. That Money was living in Los Angeles at the time of the 1857 Fort Tejon earthquake may also have heightened his seismic awareness.

Bishop, Professor, or Dr. Money—depending on his profession at the moment—was the first certifiable southern California eccentric. His oval

San Gabriel castle was guarded by two octagonal gate houses. He died in 1880, but his legacy of prophecy continued to thrive in this setting that was so conducive to natural disasters of all types.*

In the 1970s three popular books and one movie dealt with the seismic destruction of California. *The Last Days of the Late, Great State of California* by Curt Gentry was a contemporary, nonfiction look at the state from the perspective of its recent fictive demise in an earthquake. The two astronomers who wrote *The Jupiter Effect* argued that a planetary conjunction would pull the San Andreas Fault apart in 1982. Jeffrey Goodman, author of *We Are the Earthquake Generation*, cited the predictions of Edgar Cayce and Nostradamus, among others, to show there would be a major displacement of the earth around the year 2000. One of Goodman's supporting facts was that the ground was rising around Palmdale.

Throughout the first quarter of 1976 there were reports from the USGS that the ground was swelling in a huge boil centered on the San Andreas Fault in the Palmdale area. Eventually, the bulge was described as extending over a thirty-two-thousand-square-mile area from the Pacific Ocean to the Arizona border. The vast, pregnant dome was between ten and fifteen inches high.

In March of 1976, the director of the USGS sent the governor of California a letter stating that strain was accumulating and there was "increasing cause for concern over the possibility of a major earthquake in the Los Angeles area." The director flew west to meet with the governor. The USGS issued its first earthquake hazard warning.

The prediction pot was also boiling elsewhere at this time. Dr. James Witcomb of Caltech said that there would be a moderate quake in Los Angeles within one year of April 21, 1976. Later that year Henry Minturn, an amateur seismologist, predicted that a quake would strike the city on December 20. The electronic media crackled with excitement. Caltech was besieged with phone calls.

Nothing happened. UCLA's Turner later wrote of the bulge in *Waiting for Disaster:* "Two years of prior discussion of Chinese success and American

---

* Money's tradition continues to the present. As I write, the following prediction—one of many—has just been posted on the Internet: "It is our opinion that 1997 and 1998 will see an increase in both quantity and magnitude of California earthquakes, due to both the Pisces solar eclipses and the conjunction of Jupiter and Neptune." Only the medium has changed.

progress in earthquake prediction techniques made the discovery of a great-earthquake precursory sign in southern California a natural and even expected development."

All this activity aided passage of the national earthquake program. Public interest in the bulge declined rapidly after 1977. As late as 1980, the USGS was warning of a greater possibility of a major earthquake. A hazard watch was still in effect for southern California in 1981.

Again, nothing happened. There may have been no bulge; it might have been a mistake in measurement.

In January of 1981 the National Earthquake Prediction Evaluation Council—whose members, all earth scientists appointed by the director of the USGS, constituted the most eminent names in the profession—held its first meeting to consider the case of two maverick scientists who had predicted a large earthquake in Peru. The council was a club within a club. The "trial and execution" of the two scientists resembled a "court martial," said one USGS observer. Subsequently, there was no temblor.

In the most revealing and well-documented book written to date on the workings of the earthquake establishment, titled *The Politics of Earthquake Prediction*, the authors wrote of this incident:

> In future studies of scientific controversies, analysis should start, not conclude, with the bureaucratic politics of science. Indeed, if this book accomplishes nothing else, it should destroy once and for all the myth that scientists are apolitical. In the pursuit of truth but also in the pursuit of grants, contracts, research programs, travel, and prestige, scientists play politics—by necessity.

During reauthorization hearings for the Earthquake Hazards Reduction Act in 1982, the chairman of the Senate Subcommittee on Science, Technology, and Space threatened to take the program from the USGS and give it to the National Oceanic and Atmospheric Administration (NOAA). The USGS had battled NOAA's predecessor agency for the earthquake prediction program in the early 1960s and won, at least temporarily. In the meantime, the prediction budget had increased considerably, and the stakes were now higher.

Too much time and money was spent on research, and there was little to show in the way of practical results, said Senator Harrison Schmitt. A General Accounting Office report at the same time concluded that "the earthquake program has been and still is heavily oriented toward research" and the USGS would like to keep it that way.

NOAA, entrusted with the function of weather forecasting, might be a more appropriate place for the earthquake-prediction program, the senator suggested. Schmitt said he wanted a prototype earthquake-prediction system in place "within four to five years."

There were two possibilities on the immediate horizon for the USGS to choose from.

A small research effort headed by Allan G. Lindh of the Menlo Park office, who would eventually rise to the rank of chief seismologist, had been under way since the late 1970s in Parkfield. A paper published in *Science* by two other researchers in 1979 noted the regularity of Parkfield quakes. They stated: "In many respects the June 1966 sequence was a remarkably detailed repetition of the June 1934 sequence, suggesting a recurring recognizable pattern of stress and fault zone behavior." As an aside, William H. Bakun of the USGS and Thomas V. McEvilly of the University of California at Berkeley mentioned that the great 1857 quake may have been set off by a Parkfield foreshock.

In response to the senator's suggestion, the USGS also looked into establishing a prototype prediction program in southern California, since that was where the most recent action had taken place. A dense array of various instruments strung along a section of a fault line might detect early movement. The problems in southern California were threefold: which of the many faults was the most likely candidate, the high cost of such a program, and how to issue a warning for a densely populated region where the political, social, and economic consequences of a false alarm—or even the right guess—might be astronomical.

The place to plead a case and influence fellow scientists was in a refereed journal, the adjective implying impartial judgment by one's peers before publication. A strategically placed article or two were needed to validate an approach. In mid-decade, two articles on Parkfield by Lindh, Bakun, and McEvilly appeared in the *Journal of Geophysical Research* and *Science*.

Earth scientists, like other scholars, rely on self-censorship. The process excludes outsiders, a danger should public policy be involved. The vast bulk of information on earthquakes is contained in such publications. The practice has its defenders and its critics.

The editors of *Science*, who were interested in protecting their domain, once wrote:

> Peer review and editorial discretion have promoted the scientific quality of published works; the responsibility and joy of scientists in confirming or denying published conclusions have led to the eventual obscurity of bad science.

Lomnitz, an outsider, said of the peer review process in his book on earthquake prediction:

> Truth by ordeal went out with the Middle Ages. Or did it? Some day we will shudder to think that a trickle of anonymous peer opinion down our throats could have been mistaken for a fair method of reaching a scientific judgment.

Marcel C. LaFollette, a research professor of science and technology at George Washington University, wrote in her book, *Stealing into Print: Fraud, Plagiarism, and Misconduct in Scientific Publishing*:

> Americans still tend to perceive scientists as truth-seekers and as truth-tellers, but political trust has been shaken. The peer review system that validated scientific knowledge seems fallible. . . . Furthermore, the emphasis on peer review reinforces a myth that says all scientific journals use rigorous expert review in selecting all content and that the peer review process operates according to certain universal, objective, and infallible procedures, standards, and goals. Quite the opposite is true, however.

The two Parkfield articles pointed out that there had been moderate earthquakes in 1857, 1881, 1901, 1922, 1934, and 1966. The last two temblors had matching foreshocks.

The only anomaly in what "represented a strictly periodic process" was that the 1934 quake occurred ten years too early. But not to worry: "The 1934 shock should have occurred in 1944," stated one of the articles. The solution was to move it to 1944 in order to establish a near-constant, twenty-two-year cycle between quakes. A fictional 1944 event was substituted. Given such regularity "then the next characteristic Parkfield earthquake should occur in January, 1988, plus or minus 4.3 years."

On the basis of what was termed a "simple adjustment," but what more accurately should have been labeled a case of pure science fiction, the first—and so far, the last—officially endorsed U.S. earthquake prediction was issued. No one questioned it at the time.

The Parkfield experiment had other things going for it in the realm of apparent simplicity of concept. The researchers pointed out that the San Andreas Fault in this part of central California was "a relatively simple part of the North American–Pacific plate boundary," and any quake there "should guide our efforts to predict great plate-boundary earthquakes elsewhere." (Never mind the growing evidence that faults, and even portions of faults, seemed to have their own idiosyncrasies.) Then there was the chance, based on scanty evidence, that a Parkfield quake "could continue southeast along the San Andreas fault, growing into a major earthquake" that would topple Los Angeles.

The National Earthquake Prediction Evaluation Council met at the Menlo Park campus of the USGS on November 16, 1984. The chairman was Lynn R. Sykes of Columbia University's Lamont-Doherty Geological Observatory. Sykes and John R. Filson of the USGS were the first to speak. They advocated a more active prediction role for the council, according to the minutes of the meeting.

A number of sites for the experiment were discussed, including southern California, but council members soon focused on Parkfield. The talk was almost exclusively about the predictive aspect of the experiment, not the other, more research-oriented objective of "capturing" an earthquake on many instruments and then dissecting it.

Two observers spoke. Lindh distributed copies of the paper that would eventually be published in *Science*. The other article, already published, had presumably been read by all council members. Kerry Sieh of Caltech, who had formulated the trigger theory, said that the north end of the 1857 rupture zone could break in the next Parkfield shock.

Robert Wallace, a council member, took the lead in pushing for a decision on Parkfield that day. The council endorsed the prediction made by Lindh and Bakun in the draft of the 1985 article and departed the next day for a field trip to Parkfield. No one questioned the numbers. The California equivalent of the national council subsequently approved the forecast.

A press release was issued on April 5, 1985, and the public clock began ticking. The prediction specified a moderate shock occurring in the Parkfield area over a spread of eight years centering on January 1988. Such an event was assigned a probability rating of 95 percent. A USGS publication boasted: "To our knowledge, there is no other seismic zone anywhere where the time, place, and magnitude of an impending earthquake are specified as precisely."

Parkfield was sowed with the densest concentration and most sophisticated collection of earthquake-monitoring equipment in the world. There were seismographs and accelerometers and creepmeters and strainmeters and tiltmeters and piezometers. There were alignment, strong motion, dense strong motion, surface geology, and liquefaction arrays. The magnetic field, wells, telluric currents, and electromagnetic, radon, and hydrogen emissions were monitored. Two-color laser Geodimeter, geodolite, and leveling networks were established. Structures and oil pipelines were instrumented. It wasn't entirely a federal operation; the state chipped in money and installed its own instruments.

A low-level alert was issued in June 1986, after a series of tiny quakes; but nothing larger developed. There would be one hundred such minor alerts during the span of the prediction.

A few seismologists in the academic community worried that a failure at Parkfield would damage the entire prediction effort—the too-many-eggs-in-one-basket theory. The USGS was also a bit restive. Testifying before a House subcommittee shortly after the 1989 Loma Prieta quake, William Ellsworth, a research seismologist, said: "While we have focused at Parkfield with the available resources, we admit that there are many basic experiments that we are leaving undone."

By the early 1990s, it was clear to the seismological community that something had not worked.

Criticism of the Parkfield experiment first trickled to the surface in a peer-review journal. James C. Savage of the USGS called the manipulation of years "a very *ad hoc* assumption," "speculation," and "contrived" in two articles published in the *Bulletin of the Seismological Society of America* in 1991 and 1993. The second article ran under the headline "The Parkfield

Prediction Fallacy." Savage wrote that "this speculation was elevated to fact by the National Earthquake Prediction Evaluation Council." To get the two papers published, Savage later recalled, "took more than the usual amount of effort."

The outsider, Lomnitz, was less circumspect than Savage. He noted: "The easy way out was pretending that the 1934 event had actually occurred in 1944. This was ludicrous, but it was the course of action Parkfield eventually adopted." He ended his chapter on Parkfield with a quote from Thoreau: "So much for blind obedience to a blundering oracle."

Lindh would later defend the alteration of dates as a hypothesis that needed to be tested—albeit one that was "naive, simple, and had not borne up under the test of time." In an interview, he took responsibility for the idea to switch dates in the two crucial papers. Bakun and McEvilly were coauthors of one article, and Bakun and Lindh wrote the other.

LaFollette classified violations of recognized scientific publishing practices in her book. Lindh's transgression, if it was that, fell in an ambiguous category labeled "misrepresenting authentic data." She defined it as " 'the purposeful manipulation of an experimental design or results to confirm a hypothesis' or to illustrate an argument better." In some scientific fields, she said, eliminating the irrelevant and ignoring awkward facts were acceptable procedures. There was, however, a thin line dividing "allowable extrapolation" from "unethical adjustment" in these journals.

What Lindh and those who went along with him did, however, occurred far outside the domain of scientific publishing mores because it affected public policy, public expectations, and the expenditure of public funds. This is a realm where dates cannot be manipulated. The National Earthquake Prediction Evaluation Council was ultimately at fault. It should have mastered the prediction proposal, determined there was a difference between scientific conjecture and public needs, and acted accordingly.

A near-moderate quake in October of 1992 set off the experiment's only A-level alert, meaning there was a 37 percent chance of the expected temblor occurring within seventy-two hours. The earth remained silent; the crickets chirped. By the end of 1992, the expected shock had not materialized. The USGS announced its failure at a press conference at that year's December meeting of the American Geophysical Union in San Francisco.

Then a series of small shocks—perhaps precursors, it was fervently hoped—was recorded. The USGS issued another alert in mid-November of 1993. A swarm of media people and their cumbersome equipment descended on the rolling hills and oak-dotted ranch land of central California. Handsome well-dressed men and women went through the incestuous ritual of interviewing each other while they waited.

The owner of the Parkfield Cafe in the self-proclaimed "Earthquake Capital of the World" looked around and said, "There are a lot of media people. I haven't seen many geologists. I'm sure they're around, but they're probably in hiding." Once again, nothing happened. The dozen television trucks retracted their satellite dishes and departed.

At a continued cost of two million dollars a year, the USGS kept their instruments operating at Parkfield, hoping to "trap" a quake. There were two internal reviews of the Parkfield experiment; both recommended continuation and were not particularly critical. In fact, a working group of the national prediction council termed the experiment "very successful," but noted "a few problems."

As the end of the century approached, prediction was in decline and the three Cs—chance, complexity, and chaos—were in ascendancy. What Mark Twain wrote in Life on the Mississippi seemed to apply to seismology: "There is something fascinating about science. One gets such wholesome returns of conjecture out of such a trifling investment of fact."

LOMA PRIETA 8
(1989)

5:04 P.M. ON AN UNUSUAL TUESDAY

California has been extremely fortunate in terms of the timing of its urban earthquakes. The San Francisco earthquake of 1865, the Hayward quake of 1868, the 1906 San Francisco temblor, the 1971 San Fernando quake, and the 1994 Northridge event all struck in the early-morning hours, when most northern and southern Californians were either home in bed or preparing to go to work or school. Even the tsunami generated by the 1964 Alaska earthquake that devastated downtown Crescent City in the northwest corner of the state arrived in the predawn hours. During these events people were scattered about, for the most part in single-family residences.

On a normal Tuesday in the San Francisco Bay Area, 5:04 P.M. would have been the worst of all possible times for an earthquake to strike. It was a time when people would have been in large, vulnerable clumps. Crowds of workers would have been emptying into downtown streets just as glass, bricks, and marble parapets came hurtling to the sidewalks. For those already headed home, the double-decked bridges and freeways would have been clogged with traffic.

But Tuesday, October 17, 1989, was not a normal workday. The third game of the World Series was about to be played at Candlestick Park between the two local major-league baseball teams. Many workers, preceded by schoolchildren, had already arrived home and turned on television sets. Traffic on the various bridges and freeways was much lighter than usual.

A few seconds after 5:04 P.M. the earth slipped under the Santa Cruz Mountains, and the seismic waves radiated outward toward densely populated areas far from the epicenter.

Besides the vicissitudes of human activity, there were a number of natural factors that mitigated the destruction. The state was in the midst of a drought, and the many reservoirs in the Bay Area were extremely low. Had they been at normal or above normal levels, there could have been a repeat of the Saint Francis Dam disaster. The magnitude of the Loma Prieta earthquake was in the moderate range. Sixty times as much energy had been released from the shallow point of rupture just offshore of San Francisco in 1906. This time the focus was deep and was located under a dense forest sixty miles to the south of San Francisco. The period of shaking, eight to fifteen seconds, was half the expected time for the size of the quake and far less than the one minute in 1906.

Nevertheless, sixty-three people died, forty-three on the same segment of Oakland freeway. Thirty-eight hundred were injured, twenty-eight thousand structures were damaged, and more than twelve thousand people were left homeless. At a price of six billion dollars for repairs, the Loma Prieta earthquake was billed as the costliest natural disaster in the nation's history at the time.*

Lloyd Cluff, the chairman of the California Seismic Safety Commission, surveyed the debris-strewn streets of San Francisco, the broken Bay Bridge and freeways, and the ruined downtown of Santa Cruz from a helicopter and said: "I've looked at some twenty-five earthquakes in my career all over the world, and we were very lucky."

Tuesday was unusual in another way.

Sixty million television viewers in this country and millions elsewhere were transfixed by the world's first prime-time earthquake. It was the first earthquake to be almost broadcast live—almost, because the power went off at the moment of impact and it took the electronic media varying amounts of time before getting back on the air. When the various channels and sta-

---

* It wasn't. The San Francisco quake and fire left damages estimated at one-half billion in 1906 dollars, an amount worth twenty billion dollars eight decades later.

tions were finally up and running, viewers who had tuned in for a game had ringside seats at what was portrayed as a widespread catastrophe.

For the first time the reality of the event—for those who were not present—became the perception of reality gained from a small screen. The difference with written reports was that the electronic portrayal had greater immediate impact and was instantaneous. This left room for greater distortions.

The earthquake unfolded in the following manner.

The October weather was balmy. There was none of the usual fog and wind; rather, it was still, hot, and a bit sticky. Earthquake weather, some thought later.

The camera in the blimp panned over the clearly etched skyline of downtown San Francisco. Sports announcer Al Michaels intoned:

> One of the most spectacular vistas on this continent, any continent! Downtown San Francisco is the background and we zoom into Candlestick Park. For the first time in twenty-seven years a World Series game will be played in Candlestick Park. The Battle of the Bay continues.

The players were warming up, and the ABC television announcer was describing the highlights of the previous game. There was a picture of José Canseco rounding the bases, and then the image of a baseball player in his prime dissolved into visual static that resembled random patterns of tweed or the flickering hallucinations of migraine sufferers.

The first thing the spectators noticed was a low roaring noise. Then the shaking began. The glass in the luxury boxes bowed, the light towers swayed, dust rose in the air. There were scattered yells of "earthquake," then silence. When the short period of shaking ended, there were cheers and a spontaneous celebration of survival.

Ballplayers called to their families in the stands to come down onto the playing field, where they clutched each other. These gods who strut and chew and spit became mere mortals. They were scared and uncertain, like the people in the stands. The players looked childlike in their uniforms that were no longer suitable for the occasion. The erudite baseball commissioner postponed the contest, stating: "Ours is a modest little game."

As the spectators and players made their way home on the jammed freeways, they saw a column of smoke rising over the Marina District.

ON VERY SHAKY GROUND

The Marina District is a rabbit warren of million-dollar-plus homes and condominiums in the high-hundred-thousand-dollar range jammed together in an eight-by-fifteen-block area fronting on San Francisco Bay. Living in the one-half square mile area are fourteen thousand residents.

Directly on the bay are a yacht club, a marina, and a large commons used primarily for exercising and kite flying. The residential area is bounded on three other sides by the remnants of the city's military past and its commercial present: the Presidio to the west, Fort Mason to the east, and Lombard Street on the south.

In a city dominated by hills, the Marina District is distinguished by its flatness. So were other areas that were particularly vulnerable to shaking in 1989, such as the Oakland waterfront and downtown Santa Cruz and Watsonville—all distant from the point of slippage. What these areas and others around San Francisco Bay have in common is the loose soils on which they were constructed—either the alluvial deposits of rivers or the made lands that were recognized as being hazardous as far back as 1865.

The Marina is a good example of what happens when structures are built on ground that has the potential of turning to thick soup. The significance of what occurred in the Marina District was outlined by two seismologists. Thomas C. Hanks and Helmut Krawinkler wrote in the introduction to the 1991 special issue on the Loma Prieta quake published by the *Bulletin of the Seismological Society of America*:

> The Marina District holds a special place in the Loma Prieta earthquake consciousness. It is the latest chapter in San Francisco's long and sad history of recognizing the special problems of the seismic response of young Bay muds and artificial fill only after earthquakes occur, in 1865, 1868, 1906, and again in 1989. It is a reminder, too, that neither are the educated and affluent immune from the effects of earthquakes, although it remains to be seen what this politically active and powerful group of residents will extract from their local, state, and federal governments in the way of earthquake hazards abatement. Finally, the damage

and destruction in the Marina District is a case study for the profession in the great diversity and complexity of the relationships between earthquake ground motion and resulting damage.

To understand the Marina District is to be knowledgeable about its underpinnings. In California, we are what lies beneath us; and we are never sure what that consists of.

Seven miles east of the offshore San Andreas Fault, bedrock is an ancient, stream-formed basin buried two hundred to three hundred feet under the heart of the Marina. The rocks are Franciscan and serpentine, the latter being the official state rock. In northern and central California the Franciscan and serpentine formations dominate the North American Plate, while the Pacific Plate is made up of granitic and metamorphic rocks.

These basement rocks have different sources, different ages, and come from different places. The words geologists use to describe them are *assemblages, exotic terranes,* and *mélanges*. However, it was a historian who has best portrayed the bedrock dichotomy of this portion of the San Andreas fault.

By happenstance, Arthur Quinn used my home ground as the setting for his description. In his book *Broken Shore,* he compared and contrasted two peninsulas—the Point Reyes Peninsula on the Pacific and the Marin Peninsula on the North American Plate. He peered into the earth and wrote:

> Point Reyes and Marin are not merely contrary in their orientation, one to the land, the other to the sea; they differ in their very substance. The foundation of Point Reyes was laid in granite—grey granite, speckled black on white, crystalline, solid as rock should be, refined deep inside the earth, risen somehow in its purity to the surface, a thing from which monuments more lasting than human memory can be wrought. The larger Marin Peninsula has its foundation in less exalted stuff, matter in fact so common, so heterogeneous, its manner of coming to be is a deep puzzlement.
>
> Franciscan Formation is the name attached to this perplexing mixture of rocks on which the Marin Peninsula is founded. A "nightmare of rocks" one guidebook called it, a nightmare for anyone who wishes to comprehend it. And within this nightmare there is one rock characteristic of the enigma. A greenish rock, sometimes a mottled blackish-green like the bay on a clouded day, it is a kind of soapstone, a little slimy to

the touch. This greenish, slightly slimy rock is aptly named serpentine, the serpent rock. Moreover, it bears with it, like its rattles, the most arresting minerals of the whole Franciscan Formation, minerals for which there is no equivalent in the austere purity of Point Reyes: the lucent, aqueous crystals of the bluecrist; the fire red powder of cinnabar; the complex green of jade, in appearance at once solid and liquid; and even the rare manganese ores, colored like the moods of a geologist confronted with the nightmare of the Franciscan, usually a blank black, occasionally a deeply shocked pink.

Above the foundation of the Marina is a sequence of complex sedimentary deposits laid down by various natural processes over the last one million years. The sea level retreated four separate times during periods of glaciation, exposing the offshore San Andreas Fault and land as far westward as the Farallon Islands. Streams incised the slight indentation and left their sedimentary residues.

In more recent centuries a shallow cove, a tidal marsh, and streams meandering from the uplands produced bay mud, marsh, beach, and dune deposits while, in their separate times, Native American, Spanish, Mexican, and American settlers made their temporary homes by the side of the bay and came to feel the periodic shaking of what they thought was solid ground.

The jellied icing on the cake was applied in the last one hundred years, when the vertical history of the Marina District greatly accelerated.

When land became a commodity, its shape was altered by those who wished to buy and sell what became known as "real" estate. Little thought was given to the one natural force that twice within the present century temporarily altered the course of such human transactions.

The Marina District was originally known as Harbor View. The first solid structures were built just bayward of the Palace of Fine Arts in the 1860s. They consisted of a recreational complex that included baths, bathhouses, an octagonal dance pavilion, a bar and restaurant, and a shooting range. Near the end of the last century heavy industry in the form of shipbuilding, brickmaking, the manufacturing of railroad cars and various metal products, and two coal-gas manufacturing plants moved into the area and dumped their debris into the shallows.

A silver baron's grandiose plan to build an industrial park in the cove stalled after a seawall was constructed in 1893 where the current outer seawall is located. The plan was to fill in the shallow cove that lay behind the wall. Rock was hauled from San Bruno for the seawall and sand was transported from the nearby dunes to partially fill the lagoon, which now was almost completely cut off from the bay.

On the eve of the 1906 earthquake, the use of the surrounding land was partly industrial, partly residential and commercial, and partly pastoral. Brick buildings, shops, saloons, and wood-frame homes were scattered about on higher ground. Chinese tended vegetable gardens in the lower areas. A resident recalled: "After the Chinese gave up the vegetable bed, the low spaces became neighborhood dumps until finally filled in."

Damage from the great earthquake was slight and spotty, probably because nothing of consequence was built on the small amount of fill that had been deposited up to 1906. Frame buildings tilted and foundations cracked. The brick walls of one of the gas plants either collapsed or cracked, the interior wood framework was knocked out of whack, and the brick chimney stack was damaged.

Elsewhere in San Francisco, the damage on filled land was quite severe. "The unstable character of the made land on the waterfront of San Francisco has long been known," the state earthquake commission noted. Map No. 17 in the commission's 1908 report clearly delineated the filled areas: present-day Fisherman's Wharf, the Embarcadero, the foot of Market Street, and the South of Market/Mission Creek/China Basin area, where various new developments, including the Giants' baseball stadium, are planned. Twenty million cubic yards of fill were dumped along the San Francisco shoreline from 1845 to 1920.

An unknown amount of rubble from the earthquake and fire was dumped into Marina Cove. More prominent mention was made of dumping grounds in Mission Bay and off the Golden Gate. Whatever amount was unloaded, it was dwarfed by the fill deposited behind the seawall to elevate the swampy land for the 1915 Panama-Pacific International Exposition. Held ostensibly to celebrate the opening of the Panama Canal, the exposition was actually staged to demonstrate San Francisco's recovery from the earthquake and fire. Ironically, San Francisco's celebration of its rebirth set the stage for its next seismic disaster.

During six months in 1912, two giant suction dredges were stationed off the seawall and steadily pumped a carefully monitored mixture of 30 percent mud and 70 percent sand into the diminishing lagoon. From a depth of twelve feet below mean high tide, the land slowly emerged above sea level. Sand, rocks, and soil were imported from inland sites.

Into this new land and the sedimentary deposits that underlay it, more than fifteen thousand wooden piles cut into seventy-five-foot lengths were driven to provide "earthquake-proof" foundations for the many exposition buildings. At a depth of sixty feet a layer of hard green sand and clay, laid down perhaps ten thousand years ago, was mistakenly thought to be bedrock.

The 635 acres of exposition grounds were turned into a fantasyland that set a precedent for later theme parks. The walled city traversed by the Esplanade along the current Marina Boulevard was designed in a hybrid Mediterranean style, not unlike the present residential structures, albeit on a much grander scale. Elements of Moorish, Byzantine, and Romanesque architectural styles were blended in a region that depended upon the redwood tree for most of its construction needs.

A special stucco that imitated the travertine of Rome was developed to enhance the monumental quality of the buildings. The designated colors were warm Mediterranean pastels, the predominant shades of the present Marina. Pink sand was scattered on the surface of the walkways. One building, modeled on the Baths of Caracalla, was so large that it easily accommodated the world's first indoor flight of an airplane.

What went up with a great deal of planning and care came down with alacrity after the exposition closed—the process of nearly instantaneous creation and destruction being a California oddity. The more valuable materials, like electrical wiring, and a few of the structures were salvaged. Lumber that could not be sold was burned and deposited at the site, which was also opened to public dumping. The specifications for dismantling the exposition called for the piles to be cut off two feet below the surface of the ground. More mud was pumped into low areas to bring the levels of fill up to the terms of leases.

What was first a natural indentation in the shoreline of the bay became the turning basin for the shipbuilding firm, then the harbor for the exposition, and finally a small boat marina. The North Gardens of the exposition became the commons area of the present Marina Green. A spacious development named "Marina Gardens" did not materialize, and instead the land was carved up for apartments and single-family residences with twenty-five-foot frontages. In this manner, the modern Marina emerged.

A typical apartment structure built in the 1920s was a three-story wood-frame structure over a ground-level garage. The lack of support on the open first floor weakened the structure. The art deco, greco deco, streamline deco, and zigzag moderne architectural styles of the period predominated. Bay windows were de rigueur. The Marina was a much-sought-after neighborhood for those who could afford it.

Given the complexity of this vertical history, the unpredictable behavior of seismic waves, and the different tolerances of buildings to shaking, there was no way to know precisely what would occur on October 17, 1989, except that there would be grave problems—even from a quake of moderate dimensions.

The problems were caused by liquefaction and amplification. A University of California report to the mayor of San Francisco stated: "Strong (amplified) ground shaking and soil liquefaction are two separate phenomena, and, unfortunately, the Marina District appears susceptible to both." The uncompacted fill assumed the consistency of quicksand when mixed with the high water table. There were no recordings of the main shock in the Marina, but portable instruments rushed to the scene recorded motions from aftershocks that were amplified six to ten times greater than what was measured in the nearby bedrock.

History was vomited to the surface in 1989. The bits of charred wood that emerged when sand boils erupted in the Marina were either from the 1906 fire or the leavings of the exposition. Pieces of imitation travertine also rose to the surface. Startled residents discovered they were living on a post–Native American kitchen midden.

"I KNEW MY FAMILY WAS IN TROUBLE"

The human reaction to an earthquake is quite complex. Gretchen Wells was both a victim and a representative of the media. Her story, as she told it two weeks later to the California Seismic Safety Commission, incorporates elements of both roles:

I was on my way to KGO. As a matter of fact, I was outside the KGO building ready to report for duty. I was anticipating anchoring the update that evening. And my car started to shake substantially, and the whole area started to shake substantially.

I'm the mother of a new baby—relatively new. He's a year old now. And my first impulse was to go to my child. However, I did run up to the second floor of the KGO building, and I saw that there was damage. It wasn't substantial, but it was enough to prompt me to get home to the Marina.

As I was going home to the Marina there were no cars. There just was no traffic, and I was able to speed through the downtown area, very frightened for my family, but not knowing how bad it was going to be when I reached the Marina because the damage downtown obviously wasn't anywhere near the damage in the Marina. Any one of you who have seen the damage in the Marina, it's rather considerable.

And I remember getting to Bay and Fillmore and Cervantes, which is where the building came down and three died, a baby and a couple. And there was a terrible gas leak, and I looked up ahead and there was this building down. And it was just unbelievable, because I hadn't seen any damage up until that point.

And that's kind of where the Marina begins. And I knew that I was in trouble. I knew my family was in trouble, and I made my way very quickly in my vehicle toward my home, seeing these buildings just squashed all around me. It was just an incredible experience.

I cover a lot of disasters in my work. I'm a hardened newswoman, but not on a personal level. It was just extraordinary what I was seeing. Of course, the more I was seeing as I made my way quickly along, the more fear was developing within me.

And I reached the corner of Beach and Broderick, or Divisadero between Broderick—and that is where I live, and that is where the terrible destruction is, and that is where the fire began. The fire is what I would like to talk about ultimately, just the indescribable shock of being a victim, of being there in your neighborhood and seeing this just colossal chaos around you. The helplessness. The disbelief. The people standing around, just not able to move. Buildings down everywhere. The gas leaks. It was just a terrible, terrible situation.

I recall trying to make my way through the Cervantes area, and there were volunteers already taking charge. There were just one or two police officers. And, of course, they had their work cut out for them. The volunteers who came to the aid of the various authorities were just extraordinary.

But the smell of gas. I've covered a lot of gas leak stories, and this was just a very large gas leak. The danger was very evident to everybody. But no one really responded.

I want to talk about the lack of response. There was a sense that the casualty count had to be extraordinary because of these buildings that were down and because of the fires that were beginning. For about an hour there was no sense of authority. There was no help. Nobody was coming to rescue us. I guess that's kind of what we as citizens anticipate, that someone's going to come and help and rescue us.

I did make my way home finally. I remember I had to go around one person who was lying there, a man, and then there was an older woman who I had to make my way around. She was lying, and someone—there were volunteers to help. Apparently these people had just been pulled out of a building. So, I did finally reach my family. They were out on the street. We were reunited.

There were sirens throughout the city, but it was a half hour before she heard a siren in the immediate area. A lone fire truck from the nearby Presidio army base appeared on the scene. Finally, city fire trucks arrived. Gretchen Wells continued:

And you know, I'm recalling this for the first time. So forgive me if I'm a little bit hesitant. There was a large, four-unit building fully engulfed in flames, and the flames were spreading up the block. And there was no water. I want to know why there was no water.

Within minutes of the quake, nearly thirty fires broke out in the city, but the five-alarm Marina fire quickly became the most serious conflagration.

Even at this reduced level of seismic disturbance, the city nearly came unglued. Planning helped, but the random nature of what failed rendered the best-laid plans virtually useless. Improvisation became a necessity. There were good decisions and bad ones.

There was no power. Emergency generators did not start. Phones were inoperable or lines were jammed. Even if callers managed to get through, emergency operators had no firm information.

The following conversation was recorded:

Dispatcher: 911 emergency.
Caller: I need to know, did any—I just heard that something happened to the Bay Bridge.
Dispatcher: I heard that too. But I can't confirm it.
Caller: Do you know if there was anything serious? My dad is supposed to be coming home at this time.
Dispatcher: Hey, I'm not going to scare you. I don't know yet. Okay, I want you to calm down. I don't know, okay? All's I heard is that the upper deck collapsed.
Caller: (*Sharp sob and crying.*)
Dispatcher: I don't know if it's true or not, okay?
Caller: (*Crying.*) Okay, thanks a lot.

Radio communications were indecipherable. Computers failed. Fire dispatchers downtown had no idea what was going on in the streets. Fire engines idled uselessly. The smell of natural gas permeated the air. When firemen hooked hoses to hydrants, they found them dry, or the water pressure was extremely low. The fire spread. Confusion was rampant.

The regular water supply and an auxiliary system put in place after the 1906 fire for just such an emergency failed. The outdated cast-iron pipes ruptured when the soil liquefied.

One fire truck ran hoses two blocks to the lagoon at the Palace of Fine Arts and drew water from that source. Nearly two hours after the earthquake the city's lone fireboat, a near victim of a declining maritime economy and budget cuts, arrived off the Marina. It was low tide and the *Phoenix* ran aground, but fortunately it could still pump seawater through hoses to the fire. The effect was immediate and dramatic, and by 9:30 P.M. the fire was under control.

More than eighty years later some things were the same, others were different. Navy ships had also pumped water from the bay in 1906, but to no avail. Normally the wind blowing in through the Golden Gate rakes the Marina District. There was no wind that still October night, and that was why more of the city did not burn.

One of the commissioners asked if Wells had ever been told that the Marina was a hazardous place to live. She answered:

We were certainly warned, and I was aware that we were on fill. What does that mean to someone who's never experienced a 7.1 shaker? What does that mean to someone who's been living in the city for a long time and has experienced a number of small earthquakes? An earthquake was an earthquake before. Now an earthquake is something very different.

The final toll in the Marina was not the large-scale disaster that Wells had envisioned. Four residents were killed, including a baby. Four buildings were destroyed by fire and seven collapsed. Another sixty-three structures were declared unsafe for occupancy.

THE MEDIA: LIKE MOTHS TO A FLAME

The steady, candlelike column of flame rising from the Marina that first night became, along with the missing portion of the Bay Bridge and the mashed double-decked freeway in Oakland, the electronic icons of destruction for a region long thought to be susceptible to sliding into the ocean after a vast convulsion.

That apocalyptic image was in the mind of a utility company executive as he flew in a helicopter toward San Francisco:

> It was getting dark, and there was a searchlight on in front of the helicopter. Everything just looked weird. The highrise buildings looked like ghosts, dark shapes with some dim lights from emergency generator power. There was a glow over the Marina, and as we flew down into the city the sky was dark and warm and smoky and dusty. It was like flying off the end of the world.

What a difference daylight and tons of seawater dumped on roaring flames made. I had been scheduled to fly to New York the morning after the quake. Before leaving from my home north of San Francisco, I checked with the airline. Yes, flights were departing from San Francisco International Airport. That was my second surprise. The first was that I had gotten through by phone.

Armed with an evening's worth of television images, I drove over the Golden Gate Bridge with a great deal of trepidation. The light morning traffic and the thin column of smoke spiraling over the Marina were the only clues to the passage of an earthquake. At the airport the acoustic tiles that fell from the ceiling in the North Terminal had already been cleaned up.

The flight departed on schedule. A television crew met the plane in New York, and a reporter asked departing passengers in a reverent tone that suggested we were travelers arriving from the first journey into outer space: "What was it like?"

From the south a team of four media researchers from the University of Southern California drove through the night, monitoring news reports on the way. They stopped for breakfast in the Bay Area before beginning to survey the disaster coverage. Their report stated, "From the peaceful atmosphere and the light tone of conversation, one would not have known that the largest earthquake to hit the area in over eighty years had occurred less than fifteen hours previously. The contrast to media reports was startling."

There were a number of reasons for the electronic magnification of the event. The nation and the world were relatively quiet elsewhere. A large number of media people were on hand for the World Series. Disasters are expected in California, and earthquakes are associated specifically with San Francisco. What fits preconceptions is one of the chief qualifications for what makes news. The three icons—a residential neighborhood, a bridge, and a freeway—were common sights rendered uncommon, and thus very compelling, by a mysterious force of nature and the ability of long lenses to isolate dramatic images from their mundane surroundings.

Finally, as a character in Don DeLillo's novel *White Noise* points out, others enjoy seeing Californians suffer:

> "Japan is pretty good for disaster footage," Alfonse said. "India remains largely untapped. They have tremendous potential with their famines, monsoons, religious strife, train wrecks, boat sinkings et cetera. But their disasters tend to go unrecorded. Three lines in the newspaper. No film footage, no satellite hookup. This is why California is so important. We not only enjoy seeing them punished for their relaxed life-style and progressive social ideas but we know we're not missing anything. The cameras are right there. They're standing by. Nothing terrible escapes their scrutiny."

Because of speed of transmission and visual impact, the electronic media now dominate disaster reportage. Media critic Todd Gitlin, a professor of sociology at the University of California, wrote of the earthquake that "television immediately crowds into existence and reshapes it. Even victims become voyeurs." Hours later the print media provide affirmation and, it is hoped, perspective on what was seen and heard.

Radio was preeminent in the immediate aftermath of the quake. Radio stations were back on the air more quickly and the equipment radio reporters carried was less bulky than their television counterparts. Audiences associated breaking news with the all-news formats of some radio stations, and such outlets were geared to providing it. If there was no electricity, automobile and battery-powered radios were the only alternatives.

The two major San Francisco daily newspapers had no backup generators and struggled to put out skeleton editions whose news, in any case, was badly outdated by the time they hit the streets. As the power came back on and the locations of major news stories stabilized, television took over the regional market. It had never surrendered the national audience.

A former network executive described the power of the visual image in reference to the earthquake:

> But the television picture which you see is exactly how it was seen by the lens of the camera. You cannot change the impression that you got originally. You can trim around the edges of it. You can juxtapose the position of it. But you cannot change the essence of that picture. And I think as a result it leaves a sharpness of impact which does not come out in any sense from the spoken word or the written word.

Nowhere was the omnipresence of television better documented than in the two-hundred-page paperback book *Aftershocks*, published by the Marina Middle School. Again and again seventh and eighth graders recounted their experiences in the following manner: They were home watching their favorite programs, like *The Cosby Show* or *Dance Party USA*. Screens went blank. For most there was no extensive damage. They learned of devastation elsewhere from battery-powered radios and television.

For the less fortunate, there were hardships. One student wrote, "It was so boring. We had no school, no television, no power, no nothing." On their return to school, the students were greeted by television cameras. The mayor, the president, and Geraldo broadcast their separate messages from the school grounds. One student said, "All this publicity just made coming back to school fulfilling and fun."

It was not fun for everyone. Sleeplessness and television-laced nightmares haunted some students for weeks afterward. A student wrote:

> I took my remote control and turned on the TV. It was the news. After a few minutes, I got sleepy and closed my eyes. When I opened them up, I was looking straight at the Cypress freeway. I started walking toward the damaged parts when all of a sudden a little boy cried out for help. So I ran there and started digging, but no one was there.
>
> When I got up and looked again, a heavy part of the freeway collapsed on me. I cried for help, but no one was there. All that was in sight was the little boy laughing. I asked him to help me, but all he did was laugh.
>
> My mom woke me up when I was about to cry and told me it was all just a dream. In the morning I watched the news and saw the little boy who was in my dream. [Rescuers had to cut through a dead mother to get to her injured son.] I told my mom. She told me I was exaggerating, but I am sure that was him.

There were advantages to electronic images. The fire department's communications center did not realize the extent of the damage or the intensity of the flames until the blimp that was on hand to provide aerial shots of the World Series turned its camera on the Marina District. An official in the State Office of Emergency Services said, "But for the presence of the Goodyear blimp and Al Michaels [the sports announcer turned earthquake commentator] there would have been a lot of uncertainty, I think, initially."

There were disadvantages to the large media presence. The chief of highway maintenance for the California Department of Transportation (Caltrans) described the press thusly: "They're as arrogant as hell, and they're going to do anything to get in and obstruct you. And most of them are pretty nice, but some of them have no ethics, no morals at all, and you've just got to physically throw them out. It's unfortunate but true."

Because their presence was so dominant, there was a great deal of media hand-wringing after the event. Various self-appointed experts discussed the behavior of Jennings, Rather, Brokaw, and Gumbel, whose vapid electronic musings were the modern equivalent of Voltaire, Rousseau, Goethe, and Kant's search for meaning following the Lisbon quake.

A symposium was held at the Graduate School of Journalism at the University of California nearly two months after the quake. It was attended by reporters, editors, and producers who had been on the front lines. The gathering served as a catharsis. What they said also provided an excellent guide for assessing disaster coverage elsewhere, be it a hurricane, tornado, or whatever.

The city editor of the Watsonville newspaper said the media story was not about that small city with a predominantly Latino population in Santa Cruz County being ignored by the national media. Rather, she said:

> The story in Watsonville is that we lost almost ten percent of our housing and the people who were hurt the most were the people who could bear it the least, and that's the Watsonville story. What I saw on television, and I come from a different perspective, where this earthquake was a real personal thing for me, and I recognize that, but what I saw was something real turned into entertainment.
>
> We had dramatic jazzy music, we had what looked to be, if not a completely set-up situation, at least a very well rehearsed one. It was clear that here was life and tragedy and something very real turned into something false and exploitative and sensationalized. I thought to myself, no wonder everybody hates the press. Look at what we do to their stories and their lives. We turn it into the prime-time movie of the week.

Show business was very much part of the disaster coverage, particularly for those who did not call the Bay Area home. Gitlin pointed out that "no disaster is certified as great until the media heavyweights drop in." The role of the network anchors who hastily flew in from New York City and intoned their lines in front of the charred ruins of Marina District apartments or the collapsed Cypress Street freeway structure was both criticized and defended.

A former head of CBS News said their presence was "a show business trick" and "a matter of self-aggrandizement. It's promotional rather than making any real contribution to the news." A representative of ABC said,

"First of all it's a signature; and, yes, it is show business. No question about it, but I'm not going to apologize for that." An NBC producer said, "I almost would have preferred having those guys [Brokaw and Gumbel] back in the studio in New York and putting some perspective on it. It was almost like it was too hot."

Then there was the issue of the appropriateness of arriving at a scene of death and destruction in a white stretch limousine. To ABC it was a matter of having telephones in a chauffeured car, whether long or short. The CBS representative said the colors of Dan Rather's limos were not white but dark brown, dark blue, pale yellow, or dark gray.

Other journalists, it was recounted, had devised a "Rather Scale" to judge the seriousness of the particular scene. Proceeding though the day there was the blue blazer on, then off, followed by a rolling up of the sleeves, and finally the collar unbuttoned and the necktie askew. At one point Rather wore a flak jacket, someone said. Brokaw was caught by a photographer applying hair spray to keep his coif in place on the windy Marina District set.

Numbers were discussed. For the first two days the death toll was reported in the high two hundreds. The media pressured the officials, the officials guessed and were wrong, and the media ran with the number. Hardly anyone double-checked. The officials based their estimate on the number of commuters that normally drove the section of the Oakland freeway that collapsed, but October 17 was not a normal workday.

A Los Angeles reporter, who had lived in the Bay Area, remembered that the crushed lower deck carried light inbound traffic at that time of day. He cautioned his station on the numbers being released by the State Office of Emergency Services, the California Highway Patrol, and the lieutenant governor.

An Oakland reporter said, "And why is it they come up with these numbers? Because we're screaming at them." He added, "To me the exaggerated body count problem came about because of forces that I don't really understand. I had people calling me from England; and from what they saw on television, they were convinced that San Francisco had fallen into the sea with several hundred thousand people dead, not fifty or seventy."

Overseas accounts were rife with exaggeration and purple prose. "San Francisco the Damned," screamed one headline. The cables of the damaged Bay Bridge "trembled as if some invisible force was playing the harp" was another example noted at the symposium. A London paper described looting, which was almost nonexistent, in the following manner: "It was as if the quake had fractured a vat of toxic waste, spilling the horrors of the inner city everywhere." To a French publication, the moderate quake proved the whole state "could be swallowed up in the Pacific Ocean like a new Atlantis."

The number of dead dropped precipitously when a distinction was made between estimated and confirmed fatalities. A later study pointed out the conditions of "uncertainty, ambiguity, and conflicting information" that prevailed. The study failed, however, to mention the personal factor.

The vast majority of reporting chores are accomplished after the fact. Thus, some distance from the actual event is achieved, and it is easier to don the assumed cloak of objectivity. Many of the reporters, camerapeople, and producers actually felt the earth move. They were participants in a fearful event for which there was no warning and for which they were unprepared. The resident *New York Times* reporter, who had never before covered a disaster, said, "We were both reporters and earthquake victims, even if it was only psychologically."

A CBS reporter said, "What's difficult for all of us, and what puts us all on edge, is the unpredictability of it. In California there is this thing lurking out there called 'The Big One.' " There will be another earthquake, she said. "It'll be a great story. I hope I don't have to live through it, but I'd love to report it."

After the participants came the census takers. The media's performance attracted a large number of academic analysts because this country's earthquake mythology is centered on San Francisco and there hadn't been such a dramatic catastrophe for a long time. Additionally, the intensive coverage that was widely disseminated attracted attention, and the disaster took place in an accessible location.

Critics and behavioral scientists measured the media's performance by clipping, taping, interviewing, and conducting polls. For the most part, they contributed less to an understanding of the performance and effectiveness of the media than the unusually forthright comments of the participants at the symposium.

There was one exception. A team of communication researchers from the University of Southern California, who had been prepared to investigate a major West Coast catastrophe since 1986, spent two days in the Bay Area and then returned to produce their report. Their study of disaster coverage made the crucial point that the media depicted the earthquake as an isolated event. Professor Everett M. Rogers's report, published by the Institute of Behavioral Science at the University of Colorado, noted:

> The focus of media disaster reporting on events tends to attribute explanations to individual actions, while largely avoiding or underemphasizing environmental explanations and those that consider context and historical background. Thus, news media framed the Loma Prieta earthquake as a singular event, or as a series of unrelated events.

THE HIDDEN COSTS OF EARTHQUAKES

Except for the earthquake and the uncompacted soil, not much connected Hispanic Watsonville to the Anglo Marina District. In Watsonville, food was grown and processed; in the Marina, it was sold and consumed. The seismologists studied the underpinnings of the Marina District; the social scientists probed the surface of Watsonville.

The Marina received a lot of national and Bay Area media attention; Watsonville got very little. That didn't matter since Santa Cruz County newspapers and radio and television stations did a good job. Who really needed attention from New York and San Francisco, anyway?

An earthquake appears to be the ultimate democratic leveler, striking rich and poor alike. But that is not the case.

The more stable heights attract higher-income residents than the flatlands, which are more prone to shaking. (A seeming exception to this general rule is the bayside Marina; however, the more affluent Pacific Heights district overlooks the Marina.) Older rental units are not likely to have been seismically upgraded and are more easily damaged than newer homes subject to the earthquake provisions of more recent building codes. Recovery is slower because lower-income communities don't attract as much attention. Also, their residents have less political clout and lack the knowledge of which buttons to push to get action.

The initial surprise, however, is the same for all—whether expert, layperson, or official.

At the University of California campus in nearby Santa Cruz a seismologist, who had experienced large quakes, said, "I thought I was prepared for it, but I wasn't really prepared for the shock that I and my colleagues felt. Even in our laboratory I could see the symptoms." Power and computers were out in Santa Cruz. Over the hill at the western region headquarters of the USGS in Menlo Park the situation was described as "organized chaos." The telephones didn't work. Immediate communications are a necessity for scientists in these situations.

A longtime resident of Watsonville told a state commission: "Of course, we've known that we've always lived in earthquake country, and I think it was probably in the back of our minds. But we always thought it was going to be in Coalinga or Hollister or San Francisco or someplace else, not in Watsonville."

At 5:04 P.M. on October 17 the Watsonville fire chief was conducting a meeting on housing conditions, a fact he later found to be ironic:

> When the earthquake itself hit, we'd been preparing ourselves for years on how that might feel. And it wasn't anywhere near how I thought it would be. I knew from the sound of it and the feel of it that it was one of those that you didn't just start to get ready to get under the table. It was one of those that you definitely got under the table. Everybody was in that state of shock which you tend to get in when a disaster occurs. It's like an eerie feeling, like, "Now what do we do?" And instincts took over. I looked across the city and I could see the dust rising from Main Street and the columns of smoke beginning to develop throughout the city.

Liquefaction caused water and natural-gas lines to break, the electricity went out, the city's emergency generator did not work, buildings collapsed, fires ignited, panic and confusion ensued, and gradually—because, after all, this was only a moderate quake—matters were brought under control, except for housing.

An earthquake is much more than a seismic event. It races through a culture and provides a social snapshot of that brief moment. Too much attention is paid to magnitude. There are other indicators of a temblor's significance.

What was new—and what differed from the situation in the more socially static Marina District—was that the state's demographic makeup had changed dramatically since the last widespread catastrophe, that being in 1906 for northern and central California.

The representatives of national disaster-relief organizations were not prepared to deal with people from different cultures. There had been clues to the shift after temblors in the San Joaquin Valley and East Los Angeles struck earlier in the decade. Little had been done, however, to adapt to the particular disaster needs of the growing Latino population.

At a hearing following the Loma Prieta quake, a member of the California Seismic Safety Commission recalled that a representative of the Red Cross had given the same testimony concerning the organization's inability to adequately respond to the Hispanic population following the Whittier earthquake of 1987. Why was this? he asked. The representative said, "That we were still learning the same lessons over again also struck me as odd."

In Watsonville, the lack of housing and racism registered highest on the seismograph of social and political concerns. Like a sand boil, these issues erupted in the immediate aftermath of the earthquake; and their apparent resolution was the true hallmark of this event. The earthquake served as an agent for incremental social change.

Watsonville lies seventy miles to the southeast of San Francisco amid the flat, dark soil of the Pajaro Valley. Specialty crops, such as strawberries, and food processing are its dominant industries. Latinos, as Asians did before them and Mexicans before them, worked in the fields and processing plants. There was a huge influx of people in the last decade when, in the five years before the earthquake, the population exploded at a 38 percent growth rate.

All types of dislocations resulted, capped by the earthquake that put one-fourth of all Watsonville businesses out of commission. Most of the autumn harvest, held in large refrigerated storage facilities, wound up in the landfill. There was one death and massive unemployment.

The small city with a 60 percent Latino population was surrounded by an arc of Anglo affluence—Silicon Valley to the north, the seaside resorts and residential enclaves of Santa Cruz County to the west, and the Monterey Peninsula to the south. The average cost of a house in Watsonville was $150,000; but most Latinos, who had an average yearly household income of $16,000, rented quarters. The apartment vacancy rate at the time of the earthquake was less than 2 percent, and those were mostly higher-priced units.

Conditions were deplorable. Thirty to forty people were stuffed into some rental houses. Other such facilities consisted of barely remodeled chicken coops and garages. The earthquake made a terrible situation much worse. Nearly 10 percent of the housing was lost, and this consisted mostly of older and poorly constructed rental units.

Well-intentioned disaster aid providers, from the federal to the local level, were faced with a host of unforeseen problems. The victims struggled with bureaucratic procedures that would confound a college graduate who spoke fluent English and had the patience of Job.

Many quake victims spoke only Spanish; there were few bilingual aid workers. Tags affixed by city building inspectors to buildings and federal disaster applications were in English. The Red Cross established disaster centers distant from neighborhoods and within fenced grounds and seemingly solid structures. Victims wanted to be near their homes, the fences reminded them of prisons, and some had survived the recent Mexico City earthquake and did not trust rigid structures during aftershocks.

There were other cultural differences. The presence of National Guard troops and police frightened refugees from Central America, who associated uniforms with death squads. Some were illegal aliens and feared raids by the Immigration and Naturalization Service. Strange foods, like lasagna, upset stomachs.

The solution that the victims hit upon was to camp in open parks, which was the equivalent of the Chinese way. With little choice, the authorities acquiesced and provided tents, appropriate food, and medical attention. But those in charge, mostly of a different racial background, continued to pressure the campers to move to officially designated shelters, citing the unsanitary conditions and the health hazards of the impromptu tent communities.

It rained, adding to the misery of the situation. A standoff developed and became more polarized with the arrival of the media. A Hispanic candidate for city council told a state commission investigating the earthquake, "You know, it's a power struggle." The presence of the labor leader César Chávez, who arrived to lead a march for Latino earthquake victim rights, emphasized the confrontational nature of the conflict.

Earthquakes occur not only within a social context, but also in a political setting.

Latinos and the activist organizations that represented them had gone through a long, rancorous legal battle with the Anglo-dominated city government and won the right for city council elections to be held on a district rather than an at-large basis. Since registered Hispanic voters could more easily dominate a single district, this process held out the promise of getting one of their candidates elected. The earthquake forced the temporary cancellation of the election. Political distrust mingled with seismic fears. Sociologist Brenda Phillips wrote:

> Behind this election existed a history of antagonistic relations between the Anglo and Latino populations. A rapidly growing Latino community, in terms of numbers and power, brought pressure to bear upon a city which Latinos felt failed to represent them. When the earthquake struck, Latinos felt unconnected with the city's response. Latino leaders pressured the city to meet the needs of the Latino community, hundreds of whom had chosen to live in tents throughout the city. Latinos leveled charges of racism, cultural insensitivity, and exclusionary practices at the city, the Red Cross, and FEMA [Federal Emergency Management Agency].

Responding to charges of racism in the disaster-relief program, the Department of Justice investigated but could find no overt discrimination. Government lawyers suggested that relief meetings be open to Latino representatives, and this was done.

Soon cooperation and understanding replaced confrontation and antagonism, and Anglos and Latinos began to work together to find solutions. Phillips wrote, "This coalition has meant that a new Watsonville is emerging from the shadow of the earthquake. Though this phenomenon is due in large part to pre-existing conditions, the earthquake certainly sped the community toward such cooperation and social change."

The Hispanic candidate for city council won his election and was named vice mayor. For a time, the city had an ombudsman. Latinos became part of standing committees that dealt with disasters. The city sought a bilingual coordinator of emergency services and issued what could only be construed as an apology.

The city council declared, in reference to "minority groups" having been excluded: "This lack of inclusion was indeed a serious, ignorant omission from Watsonville's planning process!" The use of the phrase "minority groups" was outdated, since Latinos were now the majority.

The Red Cross, by its own admission "culturally insensitive" toward Latino victims, recruited and trained Hispanic volunteers and developed a cultural-diversity course that was given nationwide to its volunteers and staff. FEMA changed some of its burdensome procedures and became more bilingual.

The media departed to await the next disaster, which would not be long in coming. How long these changes would last and how effective they would be remained to be seen. But there was no doubt that a moderate shaking within the interior of the earth had produced some shifts on the surface.

There was, in the end, a striking similarity between the Marina District and Watsonville. In the immediate aftermath of the earthquake, officials from both jurisdictions questioned the wisdom of allowing structures to be rebuilt on soft soils. Local, state, and national recovery policies, however, were aimed at restoring the status quo.

FROM DALY CITY TO SAN JUAN BAUTISTA

The Loma Prieta portion of the San Andreas Fault extends eighty miles, from Mussel Rock, just south of San Francisco, to San Juan Bautista. This length of the fault line is doubly charged, since it contains seismic artifacts from 1906 as well as the 1989 earthquake.

From Bolinas to the north, the San Andreas Fault passes offshore of San Francisco and then rejoins the coast at Mussel Rock in San Mateo County. If there is one place where the catastrophic image of California slipping into the ocean has a measure of validity, it is along the coastal cliffs of suburban Daly City, where the fault comes ashore.

There is no sign to designate the fault's landfall in an abandoned garbage dump, whose use as a refuse pile dates back at least five hundred years. During that time the residents, be they Ohlone Indians or the latest Asian immigrants, have watched the high bluffs collapse and slide into the Pacific Ocean when the land shook. Regardless of the impermanence of the place, Daly City and its rift zone neighbors—South San Francisco, Pacifica, and San Bruno—contain the densest concentration of inhabitants living on the main branch of the fault in northern California.

The huge amphitheater formed by the San Andreas Fault at Mussel Rock, a guano-splashed black islet surrounded by smaller rocks, is reached by taking the Palmetto Avenue exit off the Highway 1 freeway in Pacifica and driving north until the road ends at the parking lot near the solid waste transfer station. Here the earth has been greatly disturbed over the years by both nature and humans.

Like a paleoseismic trench, there was a vertical order to the events that I encountered in the huge bowl on a fall day. From the parking lot I could see surf fishermen on the narrow beach two hundred feet below me. Just above the beach, a skip loader and dump truck were working in tandem to shore up the seawall that contained the garbage accumulated over a twenty-year period.

At the next level, a paraglider was preparing to launch into the northwest wind from the narrow shelf that had once been a railroad grade and then the coast highway. He was attempting to mimic the graceful flight of the circling turkey buzzards who rode the swirling air currents.

Above the birds, small planes flew north and south along the coast. At the highest level that I could discern, giant jets clawed their way upward in a series of deafening roars from nearby San Francisco International Airport.

The amphitheater is backed by massive cliffs, whose faces are either scarred an off-white color by recent landslides or are covered with ice plant that holds back the sandy soil in a state of temporary suspension somewhere near the angle of repose. The fault line tops the cliff at a point designated as Fog Gap on topographical maps. Along the crown are the repetitive facades of row houses built in the 1950s and 1960s.

There are holes in the rows where the cliff has given way over the years. Nearly twenty homes have succumbed to the forces of movement. Heavy fog, fierce winds, the noise of planes that drowns out normal conversations, older homes on tiny lots, the constant erosion of the cliffs, and the more infrequent quakes that hurry the processes of slippage have combined to keep property values down to the point where ordinary people could afford to live on the coast over the years.

At one time, the place had been inviting. There was a perennial stream and a small marsh. The Ohlone presence has been radiocarbon-dated back five hundred years by artifacts found in a kitchen midden. Archaeologists thought it was a small subgroup who used the site within the amphitheater on a seasonal basis.

The Spanish mapped the offshore rock and fault-formed basin in 1775; and the U.S. Coast and Geodetic Survey did the same in 1853, taking note of a number of sag ponds to the southeast. The Indians were decimated by diseases during the intervening years. First known as White Cliff, the area was called Mussel Rock after 1863. Irish immigrants arrived and raised potatoes, cabbage, dairy cattle, and hogs.

There were periodic attempts to utilize this unstable ground as a transportation corridor. A wealthy landowner who desired a more direct carriage route to San Francisco than the path over the top of the cliffs dug a tunnel through the rock and unconsolidated sand of the headland that ended in the offshore rocks. The four-hundred-foot tunnel did not last long, nor did the aboveground railroad or the coast highway constructed on a narrow shelf halfway up the steep slope. Landslides caused by earthquakes and erosion hastened their respective demises.

Meanwhile, the upland portion of Daly City was settled by refugees from the San Francisco earthquake. Recalling what he had seen as a young boy in 1906, a longtime resident said: "I still remember the people coming, some with a dog, a cat, or a canary in a cage. They walked out Mission Road, sometimes turning to look over their shoulders at the flames and smoke. It just seemed they couldn't get far enough away."

The farmers gave them food; the Red Cross erected tents. Real estate speculators bought land from the farmers, subdivided it, and resold small lots to the refugees, who did not want to return to the gutted city. The process of selling and reselling land with little or no thought for the natural hazards that might imperil whatever was built upon it was repeated after World War II. Henry Doelger, who had built row houses in the Sunset District of San Francisco, moved his activities south to Daly City.

What Doelger built was the West Coast equivalent of Levittown, New York, the difference being that what he developed was on or near the San Andreas Fault. The homes were appreciated by the veterans and workers who could afford them. They were derided for their lack of aesthetics, most notably by folk song composer Malvina Reynolds, who wrote a popular song about "little boxes made of ticky tacky."

The boom in housing created problems that came home to roost in future years. A history of the county states, "Lax building codes let contractors put up homes and businesses almost anywhere they wanted—on the bay flood plain, on steep hillsides prone to mud slides and washouts, and even along the San Andreas Fault."

The row houses had the same flaw as Marina District structures. The large opening for a garage on the ground floor greatly diminished the ability of the two-story structures to withstand lateral shaking. Also, the foundations consisted of unreinforced concrete slabs that extended a bare eight inches below grade. Some houses had been completed and others were under construction in 1957 when an earthquake struck on March 22. It was centered at Mussel Rock.

One person died and forty were injured in the Bay Area. The entire Pacific Palisades tract in Daly City slumped toward the ocean. Sidewalks and roads buckled. Water lines broke. Borings made after the quake revealed that the subdivision was built on a garbage dump that had been covered with sand a long time ago. Doelger made some slight modifications in the basic design and continued building homes.

Loma Prieta, whose epicenter was sixty miles to the south, caused less damage in 1989. But still the cliffs came tumbling down. Two sentences in the safety element of the city's general plan, adopted in 1994, were copied from a 1991 publication of the Association of Engineering Geologists. They read:

> Finally, it is important to note that the density of housing and other developments in the vicinity of slopes with a well documented history of seismically induced instability has increased significantly since the last major earthquake. As a result, the threat to human life and property in future earthquakes, which are only a matter of time, has also increased to the point that a disaster of major proportions is possible.

Daly City is the largest, most ethnically diverse municipality in San Mateo County. These are facts found in chamber of commerce literature that contains no mention of the San Andreas Fault.

I decided to conduct my own informal poll of seismic awareness along the streets that cling to the top of the five-hundred-foot cliffs within the rift zone. The general results were: older residents knew of the presence of the fault; newer ones, many of whom were of Asian extraction, did not.

There was one specific encounter worth mentioning. A trace from the 1906 quake crossed the lot of Art Beighley, a retired elevator mechanic who has lived on Longview Drive for thirty years.

I asked, "What was it like living on the fault line?"

He answered, "What can I say? In 1972 the *National Geographic* asked me the same question. I'm still here. You gotta live somewhere, and that's about it." He gestured toward the view of the Pacific Ocean, exceptionally clear on this day.

A few days later I found a copy of the January 1973 issue of the magazine and read the story about the San Andreas Fault. There was a photograph of a younger, unidentified Beighley in an army fatigue jacket standing before his home with his arms raised in a "What, me worry?" expression. The caption noted, "Like many other Californians who dwell along the fault, a Daly City homeowner casually dismisses any worry that a quake will strike again."

The writer asked Californians throughout the state if they worried about earthquakes. The answer he received was:

> "Earthquakes—why worry?" ran a familiar refrain. "You can get it a lot quicker out there on the freeway." "I'll take quakes to those hurricanes they get down South," I often heard. Or, "Better than those tornadoes."

I told Beighley that I could relate to what he was saying, since I lived along the fault line to the north. He continued painting the frame of a bicycle on the porch of his Doelger home, built in 1963.

I waited until the noise level from an ascending plane tapered off and then asked where he was on October 17, 1989. Beighley said he was away when the Loma Prieta quake hit. His house was not damaged. No, he had no quake insurance: too costly and too large a deductible.

Beighley was not alone. A study conducted before and after Loma Prieta, and subsequently published by the University of Chicago Press, determined that a majority of Bay Area residents "continue to eschew investment in earthquake insurance or in any of the many measures they could adopt to reduce their vulnerability to earthquake-related damage."

. . . .

I followed Skyline Boulevard southeastward along the crest of Buri Buri Ridge. The dense suburbs of the peninsula impinged upon the tree-lined parkway. From Berkshire Drive to San Bruno Avenue the major artery lies atop the fault; like other roadways in the state, it took advantage of the natural crease in the landscape.

I parked at the western end of Hillcrest Boulevard and took out my bicycle. The Sawyer Camp Trail passes through the San Andreas Valley, from which the lengthy fault derived its name. The drowned valley now contains the reservoirs that hold the water for San Francisco and the northern peninsula.

The valley was named Cañada de San Andrés by the Franciscan missionary Francisco Palóu on the feast day of Saint Andrew, November 30, 1774. In this manner, one of the world's best-known geological features gained the name of an apostle and the patron saint of fishermen, Scotland, and Russia.

The first Spanish explorers and priests favored an inland route from Monterey to San Francisco that traversed the rift valley rather than following the shoreline of San Francisco Bay. The Spanish word was corrupted to Andreas in the 1850s; for a time the dual spellings coexisted. Andrew Lawson was the first to apply the name San Andreas to the short fault he recognized in this valley.

Sawyer Camp Trail, which is paved, runs for six miles alongside, across, or on top of the fault. The popular trail crosses the crest of San Andreas Dam, built in 1869 and damaged in 1906.

From the dam's crest, the trail dips down to pass through tranquil groves of oaks. There are benches, picnic tables, and an explanation of the fault along the way. On a warm fall day, surrounded by the sparkling blue water of the reservoirs and fellow bikers, joggers, Rollerbladers, and walkers from the fitness-minded upper-middle-class suburbs of the peninsula, it was hard to imagine the cataclysmic effects of a magnitude 8 earthquake.

However, there is the potential for a Saint Francis–type dam disaster in the Bay Area. There are about sixty dams that lie within the fault system and pose a potential hazard should there be a strong earthquake. Only a few were in place in 1906. None has failed with water stored behind it, but there have been some close calls.

The Calaveras Fault, a branch of the San Andreas, passes through the foundation of Coyote Dam just east of Morgan Hill in Santa Clara County. Numerous construction problems plagued the project in 1935, and the two consulting geologists differed in their opinions on the fault and its possible effect on the dam.

Ten miles northeast of San Jose, the Calaveras Dam lies adjacent to the fault zone of the same name. The dam failed as it neared completion in 1918 and has since undergone seismically related repairs.

The Lafayette Dam in Contra Costa County is close to three major faults. The dam failed during construction in 1928, and the middle section rests on alluvial deposits.

These dams to the east were not as severely tested in 1989 as those more westerly and closer to the epicenter of the Loma Prieta quake. Fortunately, water levels were exceptionally low because of a drought, and the shock was moderate and short. Nevertheless, nine dams were damaged, two of them extensively. The two dams in the San Andreas Fault zone above Silicon Valley cracked in their first earthquake test. The consensus of seismic and dam safety experts was: "We were lucky."

The dams on Los Gatos Creek in the Santa Cruz Mountains that are closest to the epicenter are Austrian and Lexington. Two University of Santa Clara crews were on the lake behind Lexington Dam when the quake hit. As the water sloshed back and forth, they headed back to the dock as quickly as their oars could propel them. The seiche was so extreme, said a state seismic official, that "they saw the bottom of the reservoir expose itself along the margins and then fill back up quite rapidly." There were major cracks on both sides of the 200-foot-high dam, which settled nearly one foot.

The 185-foot Austrian Dam, upstream from Lexington and eight miles from the epicenter, sustained the most damage. The lake behind the reservoir was only one-tenth full at the time. Testifying before the California Seismic Safety Commission shortly after the quake, Vern Persson of the State Division of Dam Safety said: "This one settled about 2.8 feet, and it's in a very sharp canyon and the dam cracked longitudinally parallel to the crest on the upstream and downstream embankments, and it also cracked at the abutments and at the spillway structure."

He was asked what would have happened if both reservoirs had been full. Persson said, "From that standpoint we're glad it wasn't full. The construction effort on Austrian was not exemplary." Had the two dams failed like the proverbial dominoes—not an impossible scenario—the floodwaters would have surged through Los Gatos and San Jose.

I ended my bicycle ride at Lower Crystal Springs Dam, built in 1890. Ironically, the oldest dam seemed to be the safest. Only one thousand feet from the fault line, it survived the 1906 quake unscathed. A 1977 seismic stability study concluded that the possibility of failure from the maximum credible earthquake was low. The concrete-block dam was inspected after the Loma Prieta quake and no damage was visible. However, no studies I was aware of addressed the possibility of the 130-year-old San Andreas Dam failing and taking Lower Crystal Springs with it.

Lower Crystal Springs is classified as a "high hazard" structure because of the large downstream population in San Mateo and Foster City. Studies can be wrong, "maximum credible" is not a known quantity, and low does not rule out impossible for the old dam. The fragile highway bridge above the dam could fall in a quake and block the spillway. There are cracks in the dam and two intake towers. The cement is weakening.

Then there are the four large steel-and-concrete pipelines that transport Hetch Hetchy water from the Sierra Nevada to the Bay Area. They cross the Hayward Fault. One pipeline was crumpled like a beer can by the constant creep. Like Los Angeles, the San Francisco area was extremely vulnerable to the collapse of its lifelines.

The ninetieth anniversary of the 1906 earthquake was marked by three lectures given by seismologists in the Bay Area. I attended all three. While driving from Berkeley to Stanford on April 18, 1996, I passed a billboard that stated: "Risk Happens." The sign advertised the services of an insurance broker.

My route from one university to another was on Interstate 280, a scenic freeway that connects with Skyline Boulevard and parallels the fault line along the east side of the reservoirs. Just east from Interstate 280 in mid-peninsula lies Stanford University, which could have used the services of an insurance broker in the previous decade.

The university carried earthquake insurance from 1980 to 1985 but decided not to renew it when offered a policy with a $3-million-a-year premium, a $100 million deductible, and coverage for only $125 million worth of damage above the deductible. For the second time within a century, Stanford structures sustained heavy damage in 1989. There were no deaths; the dollar loss was estimated at $160 million.

There was a sense of déjà vu. Across Escondido Mall from the new earth sciences structure, where the anniversary lecture was held, stands the old geology building—officially known as Building 320, Geology Corner. It stands in the southwest corner of the original, beige sandstone quad.

Completed shortly before the 1906 event and never occupied, the turreted Victorian structure was rebuilt after the quake with the original sandstone mined in the nearby hills. The rock contained rare marine fossils, which was appropriate in terms of the subject matter to be taught in the building. Unfortunately, the masonry was not reinforced, and it was damaged in 1989.

A reporter from the university's news service accompanied Haresh Shah, chairman of the Department of Civil Engineering, as he inspected damaged buildings the day after the Loma Prieta quake. She wrote:

> From the distance one could hear the constant intense beeping of the fire alarm in the Geology Building, which had been buzzing incessantly since the earthquake.
>
> The damage on the first floor of the old building was apparent. Plaster debris was everywhere, but more alarming was an apparent space between the walls.
>
> "This is a separation of a load-carrying wall, which is not good," Shah said. And as he looked up at the ceiling to see a myriad of cracks, he remarked, "That ceiling worries me even more."
>
> A frosted skylight covers the staircase well of the Geology Building, and a macabre sight greeted Shah as he climbed to the top. Someone, evidently a while ago, had blackened in the shape of a body lying prone and near him was the shape of a dead dog. If the profile were not so stylized, one would have thought the earthquake had taken its toll.

Shah said the geology wing in the quad had sustained the most damage of the four hundred structures on campus. The reason why Stanford had a greater loss than any other institution or municipality on the peninsula, university president David Kennedy said at a news conference one week after the quake, was that it had more unreinforced masonry buildings.

As to whether the distinctive structures would be replaced with something more modern, Kennedy said, "We don't have enough history to let any of it get away." The private university was two years shy of celebrating its centennial.

Seven years later—after an infusion of federal, state, and alumni funds—the picturesque quad had been restored and, it was hoped, made more earthquake-proof. I walked through the elegant interior of the old geology building and noticed the name Branner on a plaque listing donors. The earth sciences library was named after the geology professor, department chairman, and later university president.

Up the hill from the campus, and just east of the main fault, lies the Stanford Linear Accelerator Center, run by the university for the Department of Energy. The buildings are much newer and constructed with earthquakes in mind. Materials can become radioactive during operations at the 426-acre facility; plutonium is also on hand from time to time.

The two-mile-long linear electron accelerator is one of the most precisely aligned structures in the world. It was designed to maintain a straight line that would vary less than one inch in ten thousand feet within a given year. The tunnel that houses the accelerator, located thirty-two miles from the epicenter of the Loma Prieta quake, sits in a shallow trench that passes underneath the 280 Freeway and is angled toward the fault.

Within moments of the temblor, the accelerator shifted one-quarter of an inch at a point where a known trace crossed it. Cracks appeared in the wall of the housing; the main laser had to be repositioned. There was a half inch of movement—enough to disrupt the electron beam—at another location, where no fault was previously known to have existed. It took fifteen seconds to produce the slight vertical undulations in the accelerator that would normally have taken fifteen years to materialize.

From Stanford I drove to the USGS offices in nearby Menlo Park for the last lecture of the day. One of the speakers was David P. Schwartz, the coauthor of an article that documented the poor job Branner's students did in 1906 gathering information on the San Andreas Fault from San Francisco to San Juan Bautista. Branner could not be bothered with the field work; understanding of the 1989 quake and estimating the probability of future quakes suffered accordingly.

There were two minor jolts on the San Andreas Fault that day.

. . . .

Between the university and Interstate 280 on Page Mill Road sprawls the campuslike corporate headquarters of the Hewlett-Packard Company, a pioneer Silicon Valley firm with close ties to Stanford. Both are examples of how large institutions, exemplary in their respective fields, deal with the presence of the San Andreas Fault.

The computer company also did not carry earthquake insurance, and for the same minimal-coverage-versus-cost reasons as Stanford. However, its 110 structures on the peninsula were generally newer than the university's buildings. The company's structural losses amounted to nine million dollars in 1989. There were no deaths or serious injuries, but for some it was a frightening experience.

Glass hurtled through the air in offices and large atriums and fell on desks that were fortunately unoccupied. A wedge-shaped piece of glass shot across one room and lodged in the sheetrock of the wall, much like an arrowhead. Bob Lanning, the company's earthquake preparedness program manager, told a state hearing, "If somebody had been in there, that would have been pretty serious. That suggests what the force is when the glass breaks."

The interior K-bracing failed on two buildings, which sustained major damage. In a third, the asbestos that was scheduled for removal was exposed. There were problems with the warehouse, from which two hundred million dollars' worth of personal computer products were shipped each month. Intracompany communications failed. Lanning testified:

> We weren't ready, because the earthquake program hadn't gotten in place yet. The upper-level management didn't know what to do about things. We had an emergency operations center defined. We had pulled off a drill or two in it, but we hadn't gotten the management involved yet. They didn't know what they should be doing, and they hung around. John Young [the CEO] was around for that first night and was out touring the buildings the following days.

At Stanford and Hewlett-Packard the production of goods—matriculating students at the university and electronic products at the corporation—is the first priority. The structural goal at Stanford after Loma Prieta was to have a selected group of buildings standing after a quake "that would make it possible to resume a core academic program within a targeted time of approximately one week."

An evaluation was made of all Hewlett-Packard structures. They were placed in one of three categories: C-level buildings got the least amount of attention and were constructed or retrofitted to current building codes, which meant they might take more than ninety days to reoccupy; B exceeded code requirements and could be occupied in less than ninety days; A-level structures would be given the highest level of protection and conceivably could be used within two weeks. An example of an A-level building was one that housed a high-profit operation, such as manufacturing cartridges for ink-jet printers. C-level buildings contained offices.

To the south, within the forestial landscape of the epicentral region in the Santa Cruz Mountains, the crust of the earth slipped at a depth of nearly 11 miles, almost twice the depth of the usual San Andreas spasm. The tops of trees snapped off, there were landslides, and a few houses in the remote area crumpled.

The 1989 quake did not behave in the same way as the 1906 temblor, thought by seismologists to have been the classic San Andreas event. In 1906 faulting was horizontal along 190 miles of terrain characterized mostly by a surface trace. In 1989 there was a vertical component. The surface fissures—more a product of shaking or stretching than the slip of a fault block—were scattered in confused patterns along ridgetops within 25 miles of the epicenter. Some seismologists even believed that Loma Prieta was centered on another nearby fault.

The Loma Prieta earthquake took its name from a mountain that was not the epicenter but was the most prominent landmark in the immediate area. The epicenter was south of the San Andreas Fault in the Forest of Nisene Marks State Park. The 1865 temblor that Mark Twain witnessed in San Francisco was thought to be centered in these same mountains; its magnitude was similar to Loma Prieta.

From Highway 17, the main route between San Jose and Santa Cruz, I drove southeastward on Summit Road along the crest of the ridge and made a left turn on Wrights Station Road. In the steep-sided canyon carved by

Los Gatos Creek and the San Andreas Fault I found a powerful reminder of the past.

Hidden in the dense vegetation just south of the creek was the entrance to the 6,200-foot Wright-Laurel railroad tunnel. The tunnel, completed in 1880, once connected San Jose to Santa Cruz and the more immediate stops of Wright on the north end and Laurel to the south.

Outside the tunnel two huge concrete pillars that once supported the tracks over the creek were all that remained of the railroad. The station was long gone. The north portal was deep in shade and posted with No Trespassing signs by the San Jose Water Company, owners of the upstream Austrian Dam.

The dank creek bottom was silent, except for the sound of running water. The gaping hole of the tunnel was black. From its mouth issued a small stream flowing on a bed of dirt that had accumulated along the bottom. On both sides of the portal there were concrete retaining walls that resembled small dams. They were placed there to hold back the constant pressure of a hillside periodically loosened by earthquakes and heavy rains.

I walked into the tunnel and back into tectonic history for a short distance and spoke into my tape recorder:

> It feels like Angkor Wat or a Mayan ruin. There is the residue of an ancient civilization that could construct large projects, but after a certain time could not maintain them. . . .
>
> This tunnel, and the ruins of the mission at San Juan Capistrano, are the two places along the fault line where I have seen the conjunction of human history and the natural force of the two giant plates grinding against each other. There is no question in my mind who the winner is.

The tunnel was rich in both folklore and seismic history.

It was thought to be cursed. Scores of Chinese laborers were killed and maimed in fires and explosions at the northern end during its construction in 1879. In the first week of operation an excursion train derailed, killing thirteen passengers. The 1906 quake closed the tunnel. A fracture was found about four hundred feet inside the north entrance.

Lawson asked Everett P. Carey, a San Jose high school teacher, to take a

look at the tunnel. In a slap at Branner's Stanford students, Lawson wrote
Carey: "You are the nearest resident among the people in California compe-
tent to observe such phenomena."

Carey's memorandum stated:

> The damage to the tunnel itself consisted in the caving in of over-
> head rock; the crushing in toward the center of the tunnel of the lateral
> upright timbers; and the heaving upward of the rails, due to the upward
> displacement of the underlying ties. In some instances these ties were
> broken in the middle. In general the top of the tunnel was carried north
> or northeast with reference to the bottom.

Carey noted a four-and-one-half-foot displacement near the northern
entrance. He found other fractures farther inside the tunnel. One of
Branner's students, G. A. Waring, noted a three-and-one-half-foot off-
set. Railroad engineers measured a five-foot difference. At Lawson's request,
Harry Fielding Reid of Johns Hopkins University also inspected the tunnel.
He agreed with Carey's findings.

Lawson fussed a great deal over exactly what had happened in the tunnel,
and for good reason, as it developed. Writing in 1991, David Schwartz and
Carol Prentice of the USGS said the measurements in the tunnel were the
most important of the 1906 offsets because they were crucial for making the
calculations that resulted in the forecasts for this portion of the fault.

In the late eighties the USGS had come to favor forecasts based on
percentages within a given time frame rather than the more precise predictions, such as what had been assigned to Parkfield. Weather forecasts
were the model. For example: a 40 percent chance of rain within the next
twenty-four hours; a 67 percent chance of a strong earthquake within the
next thirty years.

The two USGS seismologists thought that perhaps the Loma Prieta
earthquake did not take place on the San Andreas Fault, but rather on a
nearby thrust fault. Should this be the case the "San Andreas Fault could
still have a significant amount of accumulated stress and the possibility of
another moderate to large magnitude earthquake in this region cannot be
discounted," they wrote.

Three years after the 1906 earthquake the tunnel reopened, the narrow

gauge having been replaced by broad-gauge tracks. The tunnel was perma-
nently closed by another natural disaster—a 1940 flood. Seismic trenching
along the top of the ridge disclosed "repeated activity" in the distant past.

Because of the lack of a flashlight and general uneasiness, I remained only
a few minutes in the abandoned tunnel.

The San Andreas Fault loses its distinctiveness when it leaves the mountains
and crosses the northeastern edge of the Pajaro Valley. The fault zone again
becomes a visible presence when it emerges as the bluff on which the mission town of San Juan Bautista is perched. The bluff is known in geological
terms as a *scarp*, meaning "a line of cliffs produced by faulting or erosion."

Faulting was why the Spanish mission was founded here in 1797. There
was a commanding view, and fresh water seeped from the fractured ground.
The fault line runs along the foot of the bluff. California's first road—El
Camino Real (the royal or king's highway), which ran from Baja California
to San Francisco—followed the rift line past the mission.

There was another advantage to the site. The mangled terrain was used as
a cemetery. More than five thousand Native Americans, cajoled or forced
into service by the padres, were wrapped in their blankets and buried in the
bare ground next to the church.

There were disadvantages. In October of 1800 there was a plague
of earthquakes—as many as six a day—and the priests slept outside in
carts. Great cracks appeared in the ground and the walls of the church
were badly damaged. All the adobe houses in the settlement were rendered
uninhabitable.

The three-foot-thick adobe walls of the enlarged church were destroyed
in 1812—"the year of earthquakes." There was further seismic damage in
1865. The walls came tumbling down again in 1906, and the mission was
not restored until the mid-seventies. In 1989 there was minor damage.

On a late-October weekday I went to early-morning mass in one of the few
remaining structures in California that has a history extending back almost
two hundred years. Age has softened the vividly painted interior. Sound
reverberated in the massive chamber. A few elderly women were present.

The young priest, who served the largely Hispanic parish, announced that
on October 31, "the Day of the Dead," there would be a choice of walking to
the town cemetery or to the Indian cemetery.

A few feet from the front door of the mission church and the adjacent entrance to the Indian cemetery there was a seismograph and some information on the San Andreas Fault. The display, one of only three on the entire fault line, had been placed there in 1979 to mark the centennial of the USGS.* The information was now badly outdated.

I could not remember another place in all my travels along the fault line where the sacred, the dead, and science were fused so intimately by the convulsions of the earth.

TWO SEQUELS: NEW YORK AND NEW MADRID

Like a thermonuclear explosion, to which earthquakes are compared in terms of energy equivalents, fallout from the Loma Prieta quake spread far beyond the borders of California. Other earthquake-prone areas suddenly became more sensitive to seismicity.

A consultant's report stated that, while the interval between quakes was certainly longer on the East Coast, the ground motion for an equivalent shake in New York City was ten times greater because of the increased amplification of seismic waves in the soft soils that overlay hard rocks.

Additionally, most buildings in New York were constructed without seismic considerations or, at best, very minimal ones. New York's far greater density of population and structures, along with the height of buildings, is a catastrophe waiting to happen.

The lessons of the need for media caution and perspective were forgotten within six weeks of Loma Prieta. Given the timing so soon after the images of seismic disaster had flashed across television screens and the legacy of fear present in the central Mississippi River region, the situation was ripe for the fiasco that was about to unfold. New York City, the media capital of the nation, would play a decisive role in it.

A tragicomic circus developed around a nonevent that evoked memories of the New Madrid earthquakes of 1811 and 1812. The difference between past and present was that more than twelve million people now resided in the region that included such dense urban centers as St. Louis, Cincinnati, and Memphis.

---

*The earthquake trail at Point Reyes National Seashore and a display alongside the Sawyer Camp Trail are the other two locations.

The story broke in the Mississippi and Ohio River valleys in the last week of November 1989. Iben Browning, identified as a New Mexico climatologist, was quoted in two newspapers and a wire-service dispatch as stating there was a chance—later refined to fifty-fifty over a three-day period—of a major earthquake in the New Madrid area on December 3, 1990. He based his projection—following USGS procedures, he refused to call it a prediction—on the pull of tidal forces on the surface of the earth. It was an out-of-fashion, but not disproved, scientific concept.

The "news" spread like rippling seismic waves until it was recorded by virtually every local, regional, and national, and quite a few international, media outlets. The projected event seemed to have three things going for it: Iben Browning seemed to have excellent credentials, he seemed to have predicted the recent Loma Prieta earthquake, and he was backed by a local seismologist who seemed to know what he was talking about.

After it was shown that Browning's doctor's degree was in an unrelated field, his California prediction had been extremely vague, and the local scientist was once associated with a psychic in another failed earthquake prediction, the media deluge still continued—albeit with a few reservations.

The strongest and most courageous stand was taken by Jim Paxton, editor of the *Paducah Sun* in Kentucky, who simply halted his newspaper's coverage in late November of 1990 with the following comment:

> A national disaster does await our region on Dec. 3, but it will not be an earthquake. The nation's press has booked every hotel room within miles—not in anticipation of filming vast destruction on that day, but in anticipation of an opportunity to depict us as a region of semi-literate, hare-brained hicks (accurately, I might add, given the way things are shaping up).

What can only be termed mass madness, generated by the primal fear of earthquakes and exploited by the media, intensified elsewhere during that month. NBC set the tone with the promotion of its two-part miniseries *The Big One: The Great Los Angeles Earthquake*. In small type under the title was the phrase "A Soon-to-Be-True Story." Promos for the miniseries capitalized on the New Madrid situation. The Red Cross set up a phone bank at the Memphis NBC affiliate, and crisis counselors were on hand to calm some of the two thousand callers.

As local and regional reporters began to qualify their articles, correspondents from larger news organizations from outside the area arrived in town in search of a story. They copied what had been printed before and added new material, thus pumping new life into the nonstory.

On the front page of the November 28 issue of USA Today was a story under the headline "Fault line's threat hits fever pitch." The article implied that the quake could destroy the levee that protected the town from being inundated by the Mississippi River. The newspaper's story the previous day had quoted a resident as stating, "This is history for New Madrid."

As the countdown lessened, the coverage increased. This country's three most respected newspapers—The New York Times, The Washington Post, and the Los Angeles Times—ran page-one stories from their correspondents in New Madrid. More than fifty satellite-transmission trucks—more than had been on hand at that year's Super Bowl—were parked in town. Some came, at great expense, from as far away as San Francisco and New York City. They beamed their messages upward, and from there the "news" was distributed to the world.

As a result, panic and mass hysteria proliferated. The correspondent for The New Yorker magazine, who had the advantage of writing from hindsight, lived nearby. Sue Hubbell returned from an out-of-state trip and later wrote: "When I got to my farm, I felt as though I had stumbled into the countdown for Armageddon."

Rumors swept the area. A major utility company was said to have purchased six thousand body bags with its name imprinted on them. Mortuaries had stored extra supplies of embalming fluid. The dead would be dumped in a maximum-security prison. There were whirlpools in the river and sulfur in the water supply. The fault line was steaming; earthquake-formed Reelfoot Lake was bubbling. Blackbirds were flying backward; other birds were not singing.

A hitchhiking angel was warning motorists to avoid New Madrid. A hitchhiker got into a car, predicted an earthquake, and then disappeared. An angel was seen on one of the Mississippi River bridges. The numbers 1234567890 had special significance: 123 represented December 3; 456 stood for the time of 4:56; 78 meant 7.8 on the Richter scale; and 90 was the year 1990.

Insurance companies in Missouri did twenty-two million dollars' worth of quake business. Thousands of phone calls swamped earthquake-information centers from Memphis to St. Louis. A Memphis resident wanted to know what to do in case the lions and tigers escaped from the zoo. In St. Louis more than one thousand policemen, firemen, and emergency medical personnel participated in a mock earthquake drill. National Guard units in Missouri and Arkansas held similar exercises, and the Kentucky National Guard was called to active duty. In New Madrid and in White Bluff, Tennessee, the annual Christmas parades were canceled; and in Ridgely, Tennessee, the Christmas parade became an earthquake parade.

Religious fervor swept the region, as it had the previous century during the series of quakes. On the Sunday before "E day" churches were packed and preachers held forth from their pulpits on such topics as "Give Praise to Our God Who Can Shake the Earth If He Wants To," "Standing on Shaky Ground," and "And the Walls Came Tumbling Down." There were numerous revival meetings. Vans and cars carried such hand-painted messages on their sides as "Earthquake! or Repent!" From the megaphone in one preacher's van came the message predicting the end of the world.

Particular attention was paid to the perceived needs of youths. Schoolchildren practiced crawling under desks, as they had done decades before in anticipation of atomic attacks. Schools were closed from one to three days in five states. A Los Angeles psychologist who had experience with a quake scare in southern California arrived in New Madrid and used puppets as therapy. He counseled children to punch and kick an Iben Browning lookalike in order to convert fear to anger.

Failed prophecies were not new, and even California quake sophisticates had fallen for their share of seismic scares. Thousands fled Los Angeles in May of 1988 on the basis of a vague hint by Nostradamus, a sixteenth-century French astrologer whose obscure and rhymed prophecies were open to many interpretations. That May 10 would be the day when "the earth shall tremble" was gleaned from an Orson Welles videocassette about the astrologer, titled *The Man Who Saw Tomorrow*.

Bumper stickers declared: "Honk if you believe in Nostradamus." Sales of the video escalated, as did one-month rentals of Beverly Hills homes and the flight of frightened Cambodian refugees from Long Beach to interior valleys.

The Los Angeles psychologist billed himself as "the Nostradamus

Buster." On arriving in New Madrid, he distributed the following press release:

NEW MADRID, MO (Dec. 1, 1990)—The Nostradamus prediction of a catastrophic earthquake in May of 1988 frightened a great number of individuals in Southern California. Many were affected psychologically. Earthquake trauma phone lines, established to calm people, were jammed much of the time with hundreds of anxious callers needing reassurance. As the May 10th prediction date drew near, people became preoccupied with fantasies of destruction, hypervigilant to any sudden noise or movement, afraid to sleep indoors, with the most-jittery residents making a run on travel agents, real estate brokers and moving companies. This erratic stress behavior exhibited in California could occur along the New Madrid Fault as the December 3 prediction date approaches.

The psychologist issued a "stress alert" for the region and gave the motel phone number where he could be reached. To his credit, he did what the media failed to do. He clearly labeled it "Earthquake Prediction Hoax II— The Sequel."

With representatives of over two hundred news organizations crowding the sidewalks and streets of New Madrid and residents aiming video cameras at television cameramen, who were aiming their cameras at residents or other newspeople, a carnival atmosphere prevailed on the expected day.

The chamber of commerce sold earthquake T-shirts emblazoned with "The Great New Madrid Earthquake 1811–1812." A local restaurant featured "quakeburgers." McDonald's offered free coffee for two days. A local bar held a "Shake, Rattle and Roll" party. Wal-Mart stores in five states sold specially packaged survival items.

There was no earthquake. By Tuesday, December 4, all the television vans and newspeople had departed.

Three years later the USGS published a 250-page study that examined the responses to Browning's projection. The federal agency noted that the reservations of qualified seismologists were too late and too muted to have much effect.

The USGS reserved most of its scorn for Browning's methodology. "Mainstream science verifies its conclusions through evidence and arguments given to an audience of peer research scientists, through verification of results by independent workers, and through successful predictions based on the new conclusions." Browning fulfilled none of these requirements, the report noted, and came from the ranks of "amateurs and quasi-scientists" who make questionable predictions of natural disasters.

In fairness, it should be pointed out that Dr. Browning was seriously ill for much of this time, and he died in 1991. Browning did little to promote his forecast, nor did he need to—the media took care of that. He shunned interviews. Much of the information on his projection was gleaned from his newsletter, a video, and closed meetings with his business clients.

Nowhere in the USGS study was there any mention of the Parkfield experiment, which had run its course by the time the lengthy New Madrid report was published. It was evident that no one had a lock on prediction.

9 NORTHRIDGE
(1994)

PANGLOSSIAN L.A.

From 1857 to 1933 damaging earthquakes skirted the Los Angeles Basin. The great Owens Valley quake of 1872 was barely felt that far to the south. Unlatched doors and objects suspended from ceilings swayed in Los Angeles during the 1906 San Francisco quake. There was some damage to brick buildings and more shaking from a local quake in 1920.

The June 29, 1925, Santa Barbara temblor was felt throughout the Los Angeles area. Thirteen people died, and there was widespread damage in downtown Santa Barbara. A report in a University of California engineering publication stated: "But this low loss of life was due chiefly to the fact that the main shock occurred at 6:44 a.m., before the sidewalks and streets were filled with people, before the schools were occupied." The author of the report said that if the shock had occurred two or three hours later "the number of dead and injured very probably would have been several thousand." Bricks, stones, and heavy roof tiles from unreinforced masonry buildings tumbled to the ground in a thirty-block area.

Following the Santa Barbara quake, the people of Los Angeles were assured by their newspapers that there was nothing to worry about. The city was protected by "a great sand and gravel cushion stretching between the mountains and the sea which would absorb a considerable portion of any shock before it was communicated to the buildings." Actually, these alluvial deposits from the Los Angeles, San Gabriel, and Santa Ana Rivers amplified the shocks.

While it was true that there were older buildings, argued these commercial cheerleaders, most structures were new and were built "to provide the greatest possible protection against damage by earthquakes. There is no occasion, therefore, for anyone to be worried about the possibility of any great disaster striking the city," stated the *Pasadena Star-News*. Los Angeles newspapers echoed the same sentiments.

The local scientific community was also protective of business interests in the booming region. A 1928 publication of the Southern California Academy of Sciences stated on its cover:

> This book completely refutes the prediction of Professor Bailey Willis that Los Angeles is about to be destroyed by earthquakes. It proves that this area is not only free from the probability of severe seismic disturbances, but has the least to fear from *Acts of God* of any city under the American Flag.

Robert T. Hill was the author of the book that took Willis to task for exceeding "the province of a scientific man and an educator" by "extending the torch of alarm to interested business." As a result, Hill said, insurance rates were up and building was down.

Willis, like Hill, had worked for the USGS. He was president of the Seismological Society of America, chairman of the Department of Geology at Stanford University, and known to the public as "the earthquake professor." Willis was in Santa Barbara when the quake struck and was one of the very few geologists to have witnessed a devastating temblor. He lectured widely and was not shy about public pronouncements. A colleague at Stanford's earthquake engineering research laboratory said, "Dr. Willis was the laboratory's impresario."

Noting the long period of relative quiescence in Los Angeles, Willis wrote one year before the Santa Barbara quake: "A great shock may come soon, or within a decade, or not until after a decade. But it will come." It was that statement—so mild and prescient in retrospect—that Hill and business interests took heated exception to.

Los Angeles business interests were busy on other fronts attempting to quash the dire warnings of scientists. Henry M. Robinson, a bank president and trustee of Caltech, objected to statements by two members of the university's seismological laboratory, John P. Buwalda, who headed the lab, and Harry O. Wood, a research associate. The laboratory was funded by the Carnegie Institution.

Robinson approached Robert A. Millikan, an influential member of the Caltech faculty who was also on the seismology advisory committee to the Carnegie Institution. The banker dropped the name of the president of the Carnegie Institution. His intent was clear: get one colleague, who feared loss of funds, "to attempt to silence other colleagues," wrote Arnold J. Meltsner, a Berkeley political scientist.

Nature, in the form of an earthquake, providentially intervened, striking a note of reality in the midst of the bickering.

At 5:54 P.M. on March 10, 1933, a quake hit Long Beach. There were 120 deaths; property damage, which spread into downtown Los Angeles, was estimated at fifty million dollars. Twenty thousand homes and two thousand commercial structures sustained some type of damage.

Schools took the biggest hits. "That the loss of life was so small is partly due to the fortunate hour at which the earthquake occurred: had it come three hours earlier when the schools were in session, the dead would have been numbered by thousands," reported the National Board of Fire Underwriters.

Many of these schools dated from the 1920s, a time of great architectural extravagance and pretension in southern California. Boards of education were accused of being most interested in the size of the school buildings and their appearance. Brick buildings, particularly those that were poorly constructed and heavily ornamented, were the type of structure that suffered the most damage.

There were plus sides to the Long Beach earthquake. The state legislature passed the Field Act, which required that public schools be designed and built to resist earthquakes. Another act mandated that cities adopt building standards that were at least in conformity with the minimum requirements of the Uniform Building Code. Earthquake engineering came into being.

The Carnegie Institution established the first seismograph program for southern California at this time. A complete instrumental record of quakes in the region dates from the 1930s. Caltech took over the program in 1937 and now runs it jointly with the USGS.

The Long Beach earthquake was one of the reasons why Caltech professors Beno Gutenberg and Charles F. Richter invented the magnitude scale in the mid-1930s. They were compiling a catalog of earthquakes, and precise numbers made it possible to define an earthquake and compare it to others. Richter said:

> It was hard to persuade some persons in southern California that the destructive Long Beach earthquake of 1933 was a minor event compared to the California earthquake of 1906. This misunderstanding became serious when it was publicly argued that southern California in 1933 had "a great earthquake," that no more important shocks need be expected for many years, and that, consequently, safety precautions could be relaxed.

The journalist-historian and social critic Carey McWilliams described the degrees of seismic awareness in southern California at this time:

> On the basis of their reaction to the word *earthquake*, Californians can be divided into three classes: first, the innocent late arrivals who have never felt an earthquake but who go about avowing to all and sundry that "it must be fun"; next, those who have experienced a slight quake and should know better, but who none the less persist in propagating the fable that the San Francisco quake of 1906 was the only major upheaval the State has ever suffered; and, lastly, the victims of a real earthquake—for example, the residents of San Francisco, Santa Barbara, or, more recently, Long Beach. To these last, the word is full of terror. They are supersensitive to the slightest rattles and jars, and move uneasily whenever a heavy truck passes along the highway.

McWilliams documented the marginalization of seismic events. Local newspapers played up midwestern floods and tornadoes and said earthquakes were "not so bad." The Panglossian Los Angeles City Council passed a resolution emphasizing the goodwill generated by the 1933 quake.

Various folktales circulated in the aftermath of the Long Beach quake. McWilliams culled the following items from local newspapers:

> That an automobile on a Long Beach boulevard shook so hard during the quake that it lost all four tires; that the undertaker in Long Beach did not charge "a single penny" for the sixty or more interments following the quake; that the quake was the first manifestation of an awful curse which the Rev. Robert P. Shuler had placed on Southern California, after he failed to be elected to the United States Senate; that sailors on a vessel a mile or more offshore saw Long Beach's Signal Hill disappear from sight; that the bootleggers of Long Beach saved hundreds of lives by their public-spirited donation of large quantities of alcohol; that women showed more courage than men; that men, for some reason, simply cannot stand up to an earthquake; that the shock of the quake caused a dozen or more miscarriages in Long Beach and that an earthquake will often cause permanent menstrual irregularities; that every building not damaged by the quake was "earthquake proof" ... and that the earthquake, followed as it was by the appearance of a mighty meteor on March 24, presaged the beginning of the end.

THE SAN FERNANDO VALLEY

The region took a seismic nap for nearly forty years and then was rudely awakened when a temblor struck on February 9, 1971, along the northern edge of the San Fernando Valley—that part of Los Angeles that epitomized the good life in suburban southern California.

A report by the National Academy of Sciences stated: "The ground quaked early in the morning (about six a.m. local time) while highways were relatively free of traffic and before most workers had occupied offices in public buildings, and this minimized loss of life."

The era was post–World War II and post–Watts riot, meaning that Los Angeles was just beginning to descend the crest of the breaking wave that had made it the largest city within the nation's most populous state. It's the shape of the wave, not the size—as any surfer knows—that determines the quality of the ride. Bigness had made southern California susceptible to social and natural disasters.

I was in Los Angeles at the time. I scooped up my small son and placed him underneath me in a doorway as the cliffs of Point Fermin at the opposite end of the county from the San Fernando Valley shook beneath the creaking apartment structure. As the *Los Angeles Times* environmental writer, I subsequently participated in coverage of the disaster.

I don't recall that I had any unique insights, other than the impression that the racial conflict combined with the ruins of the earthquake had increased my sense of social and geographical instability. I had already been shaken by being present at the Watts riot and other urban conflicts, the assassination of Robert Kennedy, the war in Vietnam, and numerous wildfires and floods in California. There seemed to be no safety.

The San Fernando earthquake of 1971 ushered in the era of modern seismic history in California. The occurrence of earthquakes increased greatly, as it had before 1906. There were successes and failures in the field of seismic safety.

It was not always clear who, if anyone, could be held responsible for what went wrong. There was no knowledgeable, independent authority or effective oversight. The official reports, and the unofficial media accounts, were either written or heavily influenced by the input of the very people responsible for determining the seismology, providing structural solutions, and responding to public safety needs.

By 1971 an earthquake establishment had coalesced. The various federal, state, and local agencies; academic institutions; and private consultants molded the postmortems. No contrary voices challenged them, perhaps because what they dealt with was so truly mysterious, so conjectural, and so dominated by technical jargon that their seemingly firm grasp on matters appeared impossible to refute—if what they said could be deciphered in the first place.

Science writers should have provided independent assessments, but for the most part they did not. Their coverage was "promotional and uncritical," said Dorothy Nelkin of New York University, because they were in awe of the subject matter, idealized the practitioners, and mostly just repeated their claims.

In her penetrating book, *Selling Science*, Nelkin, who specialized in the relationship between science and the public, wrote: "Seldom do science writers analyze the distribution of scientific resources, the social and political interests that control the use of science, or the limits of science as a basis for public decisions." What was often missing in science journalism, wrote Richard Kerr of *Science* magazine, one of the most respected journalists covering the earth sciences, was "the exercise of a sort of critical judgment" that asks "how good the science is."

While schools were the chief icons of destruction in 1933, dams, freeways, and hospitals collapsed in the moderate quake of 1971. It was a dam that almost caused the unthinkable to occur.

Memories of the 1928 Saint Francis Dam disaster were briefly revived in southern California when three dams either failed or nearly gave way in Los Angeles. The first predated the quake, and should have been yet another forewarning that structural hubris was alive and well in this badly fractured landscape.

Traces of the Inglewood Fault, the same rift that had caused the Long Beach earthquake, passed underneath Baldwin Hills Reservoir in southwest Los Angeles. The minor faults were uncovered when the dam and reservoir were under construction in 1950. The Baldwin Hills facility was "regarded as a model of engineering excellence and a source of pride to its builder and owner, the Los Angeles Department of Water and Power," according to a state report on the dam's failure.

Late one Saturday morning in December of 1963 the caretaker of the dam heard the sound of running water. An alarm was given. Most downstream residents were evacuated. The initial surge of water was followed at 3:38 P.M. by "a roar like a great cannon, then a rumbling and shuddering of the ground as the face of the slope erupted," according to an eyewitness. Suburban tract homes disintegrated under the impact of the wall of water. Five people were killed and damage was widespread.

There was no earthquake. The cause of the dam's failure was more subtle. The empty reservoir revealed a crack along the bottom that merged with the break in the dam. The fissure followed the path of one of the minor earthquake faults that laced the area. Land subsiding from nearby oil drilling had slipped along the fault line.

Like the daylight drama at Baldwin Hills, luck once again prevailed in 1971. What nearly transpired could have been "a catastrophe unprecedented in this country," according to the National Academy of Sciences.

The best description is contained in a USGS fact sheet entitled "The Los Angeles Dam Story." It begins:

> Moments after the San Fernando earthquake of 1971, only a thin dirt wall stood between 80,000 people in the San Fernando Valley of southern California and 15 million tons of water poised behind a heavily damaged dam. The 142-foot-high Lower San Fernando Dam was perilously close to failure. At any moment a strong aftershock could have triggered a disaster. As it was, residents in an 11-square-mile area were forced to evacuate, while the water behind the earthen dam was lowered, a process that took 3 days.

The lower dam was only half-full at the time of the quake. A massive slide caused by liquefaction on the upstream face of the dam lowered its crest by thirty feet. A narrow, four-foot-high "shattered wall of dirt" was all that separated millions of tons of water from sleeping residents in the valley.

Like Crystal Springs just south of San Francisco and the two dams above San Jose, there was a lower and an upper dam in the southern valley. Upper San Fernando Reservoir, the southern terminus of the Los Angeles Aqueduct, was full. The earthquake pushed the crest of the upper dam five feet downstream; the dam sank three feet. The cement facing on the earth-fill dam cracked, and huge slabs were dislodged. Landslides pushed the water level higher. There were leaks; the upper dam held.

A Bureau of Reclamation report stated: "If this impoundment [the upper dam] had failed, the spill undoubtedly would have caused the overtopping and collapse of the remaining section of the lower dam." The study by the federal dam-building agency continued: "There is no question that, if conditions had been just fractionally more adverse, this event would have been recorded as one of the worst disasters in history."

The two dams, located nine miles from the epicenter, were a half century old. They had been inspected the prior year by state dam and Department of Water and Power personnel, and were judged safe. Dam safety requirements and inspections had been instituted in 1929 following the Saint Francis Dam disaster. They were subsequently upgraded after the Baldwin Hills rupture and the San Fernando quake.

• • • •

The greatest damage and most of the deaths in 1971 occurred along the northern edge of the San Fernando Valley, although the epicenter of the quake was under the San Gabriel Mountains to the north. The vertical faulting rose at an angle of thirty-five degrees and broke the surface near Sylmar.

At the time, the small thrust fault bore an unknown relationship to the San Andreas Fault, fifteen miles to the north. But what was becoming clear was that a few seconds of strong shaking caused far more damage at greater distances than direct faulting. Clarence Allen of Caltech estimated that 99 percent of the monetary loss was due to shaking, rather than the ground fracturing underneath structures.

There were fifty-eight deaths, forty-seven at the San Fernando Veterans Administration Hospital, built in 1925. That there were so many deaths in the federal hospital, which had not been designed to resist earthquakes, was hardly surprising.

What was shocking was that nearby Olive View Hospital, dedicated thirty days earlier, was extensively damaged. Again, good fortune ruled the day. The daytime staff had not yet arrived at work, and the main hospital building remained sufficiently intact so that patients could escape.

The sprawling county facility was constructed on a broad alluvial fan that spread out from the mouth of Wilson Canyon, six miles southwest of the epicenter. There were old structures as well as new ones. The latter included a two-story psychiatric unit, with administrative offices on the ground floor, and the five-story main hospital building. Both were constructed in accordance with the most recent building codes, which included seismic elements.

The second floor of the psychiatric building became the first floor. Fortunately, no one was on the ground floor. Two of the four stair towers in the main hospital building collapsed onto lower floors. Patients were evacuated down the two remaining towers. Three persons died in the main unit. Two other hospitals in the area were also heavily damaged. Four hospitals were not operational, just when they were needed most.

The ground motion in 1971 was far greater than expected. The engineering community and construction industry, who drew up the codes, had not anticipated such intense movement. In the short history of measuring ground motion, the accelerations were the greatest ever recorded.

But this should not have been a surprise, because the few accelerographs installed in southern California had recorded only one previous useful reading. That reading, used for years and still cited in the literature, was obtained from a 1940 Imperial Valley quake judged at the time—but later disproved—to be a typical moderate shake. Many a multistoried structure in southern California was designed in accordance with that relatively weak figure.

The 1971 postmortems were numerous. A federal report concluded, "One of the lessons to be learned from a study of the San Fernando earthquake is the need for more realistic approaches." A county report admitted that "the basic intent of the building code—to prevent structural collapse under high-intensity ground shaking—was not in fact fulfilled." A state report noted:

> The field of earthquake-resistant design of buildings is now and will be in the foreseeable future an "inexact" and approximate "art" or "science." It is based upon judgment gained from past experience and the relatively limited data on earthquake characteristics and effects which has been gained from our "one second" exposure to geologic time during the past 100 years or so. Damage is to be expected from future earthquakes.

While schools, dams, and hospitals were traditional structures, freeways were a more recent marvel. The graceful, multilayered interchanges became, along with the palm tree, the most prominent symbol of Los Angeles. They represented freedom. Their star would dim. Gridlock, random shootings, and knowledge of the dire social consequences of freeways came later.

Driving the freeways, as novelist Joan Didion made plain in her 1970 book *Play It As It Lays*, was an experience in extended liquidity. The noted architectural critic Reyner Banham called one freeway interchange "a work of art, both as pattern on the map, as a monument against the sky, and as a kinetic experience as one sweeps through it." His book *Los Angeles: The Architecture of Four Ecologies* just barely predated the San Fernando earthquake. In that widely acclaimed book, there is no mention of earthquakes, and the citation for freeways commands the most words in the index.

These arched bands of concrete delicately balanced on thin pedestals had not yet been tested. When they were in 1971, two major interchanges at the north end of the valley were rubble. In all, approximately sixty highway and freeway bridges were damaged. One-fourth of them either collapsed or were severely impaired. One-half were moderately damaged. Most of these forty-five bridges were in a five-mile-long corridor a few miles southwest of the epicenter.

The greatest damage was sustained at the Golden State and Foothill Freeway interchange, which was under construction and 95 percent complete at the time. There was also extensive damage at the nearby interchange of the Golden State and Antelope Valley Freeways, also being built with the latest seismic precautions incorporated in the design. Three people were killed, a minuscule number considering the amount of traffic that would have been on the freeways an hour or two later.

Both interchanges were on the major north-south route that crossed Tejon Pass. At the time, nearly one hundred thousand vehicles a day traveled the Golden State Freeway. It would be months before traffic would again move relatively smoothly.

The world-famous California freeway system suffered a mortal blow at the very apex of its development. The National Academy of Sciences concluded: "Present standard code requirements for earthquake design of highway bridges in high-risk areas are grossly inadequate and should be revised."

THE PACE QUICKENS

There were earthquakes elsewhere in the West.

Very few people have seen fissures opening in the earth, although such a phenomenon is one of the most frequently imagined calamities. Two elk hunters were driving on a dirt road in Idaho when the Borah Peak earthquake—the strongest in the state's short history—struck in the fall of 1983. Their vivid description of an earthquake is one of the best that exists. The driver, Don Hendricksen, said:

> Okay, I got about to this point right here, and all of a sudden I did feel light-headed and I just lost my equilibrium. I felt like I was going to pass out. I was just about ready to tell John at that time, "There's something wrong with me, John." And soon after that, right after that, it just started shaking like crazy. The Bronco was off the ground completely, it

was just rocking like this. And right soon after that the bank dropped, and I was hanging on to the steering wheel. I looked over to John, and he was just flying in the passenger seat. He was between the two seats and trying to get up. And he was saying, "What's going on?" Or something like that; it was to that effect. I wasn't about to answer because I didn't know. I was looking out the window, just looking all around not knowing what was going on. I looked out the window and this ground right to my left here was opening and closing, and I was really terrified at that. I figured it was going to open up and drop us right in. And it all happened, I don't know, probably in a span of ten or fifteen seconds.

The geologists who interviewed them said that, although the two men had no formal training in the earth sciences, "yet they provided a remarkable account of what must have been a very frightening experience."

East Los Angeles and the San Gabriel Valley were next in California. The Whittier Narrows quake of 1987 caused less than a dozen deaths, and some ten thousand structures sustained three hundred million dollars in damages. These were indicators of a relatively minor quake. The shock struck at 7:42 A.M., causing some media outlets to speculate that the Los Angeles area was prone to early-morning temblors. The term "wake-up call" began appearing in the literature of earthquakes.

For the first time social and psychological factors became part of the seismic equation, as they would two years later in Watsonville. The area hit hardest had a high percentage of Asian and Hispanic residents. They responded differently.

The emotional toll was given some attention for the first time. A report stated:

> Adults complained typically of frayed nerves, inability to sleep, depression, fatigue, headaches, nausea, loss of appetite, and anger. Children responded with symptoms including clinging to parents, nightmares, screaming in the night, bedwetting, headaches, fear of sleeping alone, fear of separation from the family, and physical complaints. Fear of going back into homes with even minor damage was common.

The early-morning Landers earthquake of June 28, 1992, was notable for not being on the San Andreas Fault, for not being in the Los Angeles Basin, and for being quite large. The ground ruptured for a distance of 50 miles in the Mojave Desert, just northeast of Palm Springs. It was the longest break since 1906. Landers released more than twice the energy of Loma Prieta.

But even more remarkable, the Landers quake set off seismic activity as far away as Yellowstone National Park, 750 miles to the northeast. Faults thought to be separate were now determined to be interconnected. There was one death in the desert. Strong aftershocks reverberated throughout the heavily populated coastal areas and bred fear of malarial proportions.

All that was needed was for the seismologists to say that "the Big One" was on its way—and they did just that. Accordingly, rumors proliferated. A volcano was thought to be erupting and spewing lava near the San Andreas Fault. The water table was dropping, a sure sign. Caltech had ordered its employees to flee, a variation on the standard rumor that scientists knew the time of earthquakes but kept that information to themselves. A woman wrote a female USGS seismologist: "I know you can't tell me when the next earthquake will be, but will you tell me when your children go to visit out-of-town relatives?"

Seismologists noted that the rate of quakes was increasing in southern California. A working group on probabilities stated:

> Although southern California lacks an adequate historical perspective with which to consider the significance of this increase in seismicity, it is interesting to note that in the fifty years preceding the great 1906 San Francisco earthquake, the rate of occurrence of moderate-sized earthquakes in northern California was significantly greater than for the following several decades.

A year and one-half later, on January 17, 1994, another moderate earthquake struck the San Fernando Valley. The Northridge quake arrived at 4:31 A.M. on the day set aside to observe the birthday of Martin Luther King Jr. Given the hour and the holiday, good luck was once more abroad in the land.

Again, the main thrust of the quake was felt along the northern edge of the valley—more a suburb that lies within the sprawling city limits of Los Angeles than a true urban environment. Indeed, it could be said that no large quake has ever struck the densely populated core of a modern California city.

There were the familiar images of broken concrete, twisted rebar, deflated arches, flat sections of bridges skewered by their supports like so many barbecue items, crushed vehicles, cars teetering halfway off the edge of nowhere, deep voids where there had once been the assurance of safe passage, and massive sections of concrete blocking roadways.

Most remarkably, there was silence and no traffic along these ghost highways—once the very symbol of a vital state. Once again commuters took alternative routes that were crowded, or they lined up for other means of transportation until the fragile spatial links were reestablished.

By now there was a discernible pattern: moderate quake, short period of shaking, record high ground motions, and no warning on a previously unknown fault; over sixty dead, thousands injured, emotionally crippled, and homeless; billions upon billions of dollars' worth of damages; sporadic fires, the failure of the water system, and fortunately no wind to fan the flames; and structures that should have survived unscathed did, others failed, and there were surprises.

Cracks were discovered in the welds of the steel-frames in high-rise buildings. There were fears they might collapse in another quake. Engineers were alarmed. Astoundingly, it was determined that there had been mixed results in the testing of such a key construction feature; and the tests themselves were flawed. A state report noted, "There is a vast inventory of steel-frame buildings throughout the state (and the nation) that use details of construction similar to those that failed."

Three years later what caused the cracks had still not been determined. Because of the problem with steel welds and the far greater displacements expected to accompany a magnitude 8 quake, Caltech's Tom Heaton—the Bailey Willis of his time—suggested in 1997 that California should limit the height of buildings to ten stories. That same year the chairman of the California Seismic Safety Commission said there were forty-four buildings in San Francisco and Oakland that had faulty steel welds dating from the Loma Prieta quake.

The economics of land use led to death in wood-frame buildings. Apartments and condominiums with garages on the ground floor were damaged or collapsed in excessive numbers. Sixteen died when the Northridge Meadows apartment structure lost a floor. A report of the Earthquake Engineering Research Institute said:

> Originally conceived as a single-family-house environment, Los Angeles eventually began to be squeezed for land, and housing densities had to increase. Beginning in the 1950s, the low end of the Los Angeles housing market was addressed in volume by the low-rise apartment house, initially a two- or three-story wood-frame minimal stucco box, often with all or some of the second floor supported on pipe columns to permit the garaging of the essential automobile. Thus the "soft story" that damaged or destroyed so many of these buildings in the Northridge earthquake had its origins in functional necessity.

In terms of structural safety, southern California led the nation in seismic requirements. Yet there were any number of ways a poorly designed and constructed building could become a reality. Most had to do with cutting costs. They were standardized engineering solutions that might not be appropriate, no continuing education requirement for certified engineers, poor plan-review practices by local jurisdictions, shoddy construction, little or no building-site inspection by the engineer, and building inspectors who were not knowledgeable or were inept.

The noted earthquake engineer Henry Degenkolb had said of the state licensing board: "The board has not been effective in weeding out incompetent engineers. Once an engineer gets his license, if the guy doesn't make waves, he can get by without doing very much." As for continuing education, Degenkolb said, "For a lot of engineers it's more interesting to go skiing or whatever, than keep up."

Another surprise was the potential lethal effects of nonstructural damage. Heavy bookshelves, electronic equipment, cabinets, ceiling systems, and light fixtures crashed to the floor in vacant offices and schools. A state report noted what could have occurred had schools been occupied: "Even if the children had 'ducked and covered' under their desks, many would have received serious, even life-threatening, injuries from falling and breaking furnishings and equipment."

There were less tangible side effects. Dust from the many landslides in the nearby mountains caused an outbreak of valley fever, a respiratory infection brought on by inhaling spores of a fungus present in soil. Three people later died of the disease.

Terror was widespread. At a March congressional hearing a subcommittee member, Jane Harman, related a typical reaction:

> I live in Marina del Rey, California, and my husband and one of our children huddled under a doorway while the glass shattered and the plaster fell off the walls. We were lucky because the damage to our home and most homes in my congressional district was fairly modest. But it was thoroughly terrifying—the most terrifying California earthquake I have been through.

The emotional toll of the earthquake became part of the mix of the region's general malaise produced by the racial conflict that broke out after the first Rodney King trial, frequent wildfires and floods, and a deep recession. There were stories of people fleeing the state, which had experienced thirteen federally declared disasters between 1989 and 1994. An engineering report on the quake commented, "These events helped create a widely held impression, reinforced by media reporting, that Southern California was more or less one continuous disaster area."

Testifying before the House subcommittee, Kaye M. Shedlock, the head of the USGS's earthquake branch, said, "We really don't know what an 8 is going to do. But the previous experience shows us that we have been underestimating consistently all along." She added, "I think an 8 is going to be a real stunner for all of us."

A subcommittee member asked what would happen in a really large quake. Caltech's Heaton replied: "To be honest, as was pointed out earlier today, we really don't have a very good answer to that question, and that is pretty disquieting for those of us that live in the area." Heaton said of the moderate series of quakes: "When the history books are written, I think they will get a footnote."

Blame needed to be placed for immediate events, however, and it eventually was in a *Los Angeles Times* publication:

> The tough part about an earthquake is that there is no one to blame. Building inspectors and freeway engineers and stingy landlords make

unsatisfactory patsies. The real "culprit" is nature, and what's the use of complaining about that.

A new seismic term gained prominence in the mid-1990s. Blind thrust faults, such as the Northridge fissure, became the center of attention. The phrase held a note of menace. In this land of celluloid concoctions, it conjured up an image of a King Kong–type creature lashing out indiscriminately.*

There were more than three hundred faults underneath southern California, some long and visible and others short and imperceptible. One blind thrust fault passed directly through downtown Los Angeles; another bisected the Wilshire District. Their angled, vertical motion was inordinately destructive and unpredictable. The visible San Andreas Fault to the north seemed relatively benign by comparison.

The best guess was that the northwest-moving Pacific Plate was bumping against the east-west-trending Transverse Range portion of the San Andreas Fault, and the Los Angeles area was being gradually compressed. Unseen wrinkles were forming within the folds of the earth's skin. When a sneeze contorted the creature's face, the lines suddenly became visible.

I began to look for an opportunity to descend into the belly of the seismic beast.

INTO THE BELLY OF THE SEISMIC BEAST

Ever since watching Superman in the form of actor Christopher Reeve fly through the San Andreas Fault, I had imagined walking underground through a fissure in the earth. So when someone on an Internet news group devoted to earthquakes mentioned that the people who were digging a subway under the Santa Monica Mountains had encountered the Hollywood Fault, I immediately got on the phone with a representative of the Metropolitan Transit Authority (MTA) and arranged a walkabout.

That was how I came to be standing in the parking lot of the Fifth Church of Christ, Scientist, at the corner of Hollywood Boulevard and La Brea Avenue. A banner with a Mary Baker Eddy aphorism was prominently

*Indeed, such an angered beast of cyclopean dimensions was depicted at the Universal City entertainment complex.

displayed in front of the church. It read: "Divine Love always has met and always will meet every human need." I fervently hoped that was the case as I prepared for my descent.

Waiting for me in the parking lot rented from the church was Stuart Warren, geotechnical manager on the project that consisted of pushing twin subway tubes through the Hollywood Hills, and Mary Ann Maskery from the MTA's media-relations office. Stuart was a veteran of many such undertakings, beginning with coal mines and proceeding through the tunnel under the English Channel. Mary Ann was my constant shadow at interviews. This would be her first trip underground.

I had been outfitted at the Universal City construction site with rubber boots, white coveralls, an orange vest, a hard hat, safety goggles, a flashlight, a breathing device should carbon monoxide be encountered, and two brass disks. One would be left at the top of the elevator and would indicate that I was missing, should there be an accident in the tunnel. The other would be inserted in my body bag when I was brought to the surface. I signed the obligatory release form.

Looking like a comedic astronaut in all my paraphernalia, I clumped across the parking lot with the others and climbed the wooden steps to the Hollywood construction site office. We entered the antechamber to a world far removed from the tourists who were just beginning their daily promenade along the nearby Walk of Fame on Hollywood Boulevard.

During interviews with engineers and geologists, the safety of all phases of the project, including its eventual operation, had been stressed. Yet I couldn't stop thinking that these were the same folks who were responsible for sinkholes, large cost overruns, and dreadful mismanagement. Some of the problems were endemic to digging a 2.4-mile-long tunnel under thousands of people and their artifacts. Better to dig a hole under the English Channel, or the desert.

The region's biggest and most maligned public works project did seem ill-fated, however. "Litany of woes" and "trouble-plagued" were some of the standard phrases applied to the extension of the Red Line from Hollywood to Universal City. Not encouraging, I thought.

Optimism, in the form of Bob Edwards, a tunnel inspector, flooded the small office. Tunnels were safer than anything built upon the surface of the earth, I was told.

"Ten thousand feet up in the Sierra Nevada there was a 5.5, and I came out of the tunnel and didn't even know it," said Edwards. He added, "In San Francisco two years ago, I was in a little one at the end of the Muni turn-around on Market Street. I didn't feel a thing." Others in the room recalled their benign seismic experiences. It was a virtual Greek chorus of assurances.

I wasn't as big a rube as I may have seemed to these veteran tunnelers. Most books in the underground portion of the main Los Angeles library stayed on shelves, while those aboveground tumbled to the floor in the 1994 quake. The coal mines under Tangshan, China, were relatively undisturbed in 1976.

There is, however, a wide gulf between the reality of safety and the perception of a particular situation—witness the statistics of flight safety versus the irrational fear of flying. We all have our phobias. Mine happens to be heights, not depths, which was fortunate for me on this day.

As if divining my thoughts, Stuart pointed to the plans for the 300-foot Special Seismic Section of the tunnel. It was built wider through the 120-foot active fault zone in order to facilitate repairs, should there be a damaging quake. Flexible steel girders will line the walls, which will also be covered with layers of shotcrete—sprayed concrete with steel fibers. Fifteen-foot bolts—soil nails, in effect—will be embedded in the crushed rock.

I walked outside and looked down upon the cramped construction site that surrounded the gaping hole in the earth. A rectangular sound baffle composed of blue, green, and gray panels separated the construction activity from the surrounding apartment dwellings. Inside the enclosure, dump trucks backed up to receive their allotment of primordial ooze. They were "mucking out." Fifty-two truckloads of muck had been hauled away that week.

As we approached the hole a crippling burst of fear imploded inside my stomach. The steel grating we had to cross passed over the 116-foot-deep black hole. I breathed deeply and frequently, a partial solution that I've employed over the years to deal with heights. I willed myself forward and muttered: "Don't look down!"

We crossed to the elevator. I gingerly waited on the exposed grating for its arrival while Stuart blithely explained the brass check procedure. "It's in case you get squashed, or something like that," he said.

We descended into the pit. At the bottom was a sign: "Danger: Laser Light." There was, in fact, danger all around. Poisonous carbon monoxide spewed from the exhausts of heavy machinery. Deadly hydrogen sulfide gas and fire-prone methane could seep from oil deposits, if encountered. There was also the ever-present danger of cave-ins.

A panel of consultants had noted: "Problems experienced on previous tunnels [in Los Angeles] included boulders, caving, sinkholes, methane, as well as an explosion and deaths which occurred in 1971." The panel thought the planning and design for the Hollywood Hills tunnels had taken these dangers into account.

We were at a depth where the fossil remains of ancient redwood and pine forests predated the 46,000-year limit of a carbon-14 dating technique. It was here that the bones and teeth of sloths, camels, bison, and elephants— the remnants of a past ice age—had been recovered by tunnelers. This was the prehistoric graveyard of Los Angeles.

The tunnel was warm; there was almost a jungle type of humidity more than one hundred feet below Hollywood. That surprised me. Other tunnels and caves that I have visited were cool, which was why I wore a sweater under the coveralls.

I began to sweat. I asked Stuart about the temperature. He said something about it being southern California, ha-ha.

A slight mist dimmed outlines in the tunnel and gave the soft-focus scene a feeling of unreality. The evenly spaced lights on the walls cast a dull glow. The noise of the huge fan that sucked out the foul air was deafening. Altogether, there was a complex assault on the senses.

With the current spate of natural-disaster films in mind, I remarked that the tunnel would make a perfect setting for a movie. I could imagine the presence of monsters, alien beings, and chase scenes. It oozed catastrophe.

Stuart looked at me but did not respond. I imagined him thinking: "That's all I need."

Now began the slog through the tunnel, which curved gently northward. Stuart set off at a fast pace, as if to distance himself from the problems that we represented. I followed a short distance behind. We stopped a couple of times to let Mary Ann catch up. She was a quiet presence; I had no idea what she thought.

The stygian setting was the kingdom of the tunnelers. Men were lone, silent figures who toiled as outlines, not precise shapes. They appeared, then vanished mysteriously, working at a rhythm and in a place that were quite different from the rhythms and habitats of ordinary mortals. Stuart fit in easily; we didn't.

Besides being hot, my boots were uncomfortable. I could feel a blister or two developing. Walking was quite tricky. The bottom of the near-circular tunnel was covered with a thick coating of mud and water. Two steel rails barely protruded through the muck, forming additional obstacles.

Like the heaving deck of a ship, the canted surfaces offered no sure footing. I felt as though I might fall at any moment, land in the thick goo, bang my head on a steel rail, and be carried out in a body bag. A sense of security was lacking, on many levels. Stuart's figure kept retreating; I hastened to catch up.

We plodded through the alluvial deposits from the Los Angeles River that were bookmarks at either end of the tunnel's geology. Stuart picked up a handful of mud and described its properties: "Sandy silt, pale to medium brown, pretty nondescript."

Precast concrete sections lined the tunnel through the alluvium. Spray-painted on the concrete was the graffiti of the diggers. "Graveyard [meaning shift, not cemetery] was here" and "We'll miss you hard-headed hippie" were two examples that would eventually be covered by another lining. The esprit de corps of tunnelers seemed akin to that of submariners.

I could see amber-colored lights in the distance. A widening indicated the Special Seismic Zone. The concrete siding ended and was replaced by a coating of gray shotcrete—a stuccolike substance that formed a thin covering over the rough, bare walls of the newly dug tunnel. The walls leaked water; puslike excretions seeped through the shotcrete.

It was an ideal time to walk underground through the fault. One tunnel had just pierced it completely, and the other was halfway through. I had hoped that the raw earth of the fault might be exposed, but Stuart demonstrated why it had to be covered immediately. He broke off a small section of the shotcrete and dug out a piece of the fault gouge from behind the thin covering. "See," he said, "this is quite gritty. It is a very, very shattered material. Quite weak."

Before digging through the rift zone they had sunk a horizontal bore through the fault line to determine what lay ahead. I had seen the corings laid out in stacked boxes at the Universal City office. The fault, Stuart explained, as we stood within the actual fracture zone, was not a single band but rather "a multitude of individual planes and shears." He sketched its complexity for me; it looked nothing like the fault that Superman had traversed.

I needed a few moments to myself and walked a short distance away. I wanted to gather my thoughts and fix the scene in my mind.

We were alone. There were no workers at the twin ends of the tunnels today. The activity was elsewhere. Three yellow digging machines were parked at the end of the right tunnel, right and left being determined from where one stood at Union Station, the epicenter of the underground transportation system.

First the top half and then the bottom half of the tunnel was excavated. A steel brace supported the cyclopean eye of a laser navigation system at the end of the tubes. The same procedure was used by the tunnelers coming from Universal City. If all went according to plan, they would meet at an undetermined date in the near future. (This was subsequently accomplished.)

Water was seeping from the end of the tunnel, the rawest wound in the shattered earth. The minerals limonite and calcium carbonate were mixed in the water; the former left an ocher stain, the latter, white marks. What looked like white fluffy popcorn was a leachate material from the shotcrete. It all seemed to me like the bitter residue of Hollywood.

In terms of geologic time, the subway would have a very short life. The constantly shifting earth under L.A. would soon reclaim this slight intrusion and grind it up and spit out the concrete and steel and shotcrete as fault gouge, perhaps to confound some geologist's expectations in the distant future.

On the surface, the Hollywood Fault is a distinct entity that traverses one of the best-known landscapes in the world. Mention Hollywood, the Sunset Strip, or Beverly Hills and very few people would guess that an earthquake fault united them.

If an X ray could be taken of the crust of the earth underneath the Los Angeles Basin, it would resemble a car window smashed by a baseball bat. One of the more distinct cracks would be the Hollywood Fault, extending from the Los Angeles River near Griffith Park to a point near the Beverly Hills Hotel. Truly, no other seismic rift in the world is more star-studded. It is also a study in social contrasts and monetary extremes.

The Hollywood Fault is a known fracture, of which there may be some one hundred scattered about the Los Angeles Basin. There are more blind thrust faults, of which only a few—like the Northridge Fault—have announced their presences during historic times.

Kerry Sieh of Caltech was the geological consultant for the tunnel project, and his opinions were constantly cited by the engineers. Sieh determined that earthquakes might occur on the Hollywood Fault, on average, every 1,000 to 2,500 years. Besides the fact that quakes do not occur at regular intervals, Sieh did not know the date of the last quake. Thus, there was no telling when the next one would strike.

Given a spread of a thousand or more years, it was no wonder that a tour of the fault line from Hollywood to Beverly Hills revealed almost total ignorance of the presence of such a potentially disruptive feature. My guide was a technical paper written by James F. Dolan of the University of Southern California, who has also conducted field trips along the fault line.

I picked up the fissure at Beachwood Drive and Franklin Avenue, directly south of the HOLLYWOOD sign. This was "Now Renting Weekly" country. West of the Hollywood Freeway the fault split into three parallel strands: along and just south of Franklin Avenue, north of Franklin, and the Yucca Street strand. Behind the landmark Capitol Records tower, the fault scarps on Franklin and Yucca Streets were quite pronounced.

The windows were boarded up and there was a fence around an unreinforced masonry apartment building, a victim of the 1994 quake. These were the mean backstreets of Hollywood.

Echoes of apocalypse hung in the air, this being the locale of Nathanael West's *The Day of the Locust*. West also wrote the novel in this neighborhood.

Three years before Northridge, Lionel Rolfe described the area in his book, *In Search of Literary L.A.*: "Many of the apartment buildings are unreinforced masonry, doomed to extinction when the inevitable big earthquake strikes. The orgy/riot climax of West's novel seems metaphorically to pinpoint this madness."

I walked down to Hollywood Boulevard and had a cappuccino in a mini-mall off the main thoroughfare. There was a tourist information center run by the Los Angeles Convention and Visitors Bureau. I asked the man behind the counter where I could find the Hollywood Fault. He said he didn't know. "It's not an attraction," he added.

Farther west there was an attraction. On Franklin Street, just behind Mann's Chinese Theater, a sharp dip in the terrain at Hollywood Franklin Park announced the continuing presence of the fault line.

At nearby Runyon Park, the trails were dominated by morning dog traffic. People were walking their pets in brilliant sunshine. Metal bowls were affixed to chains attached to water fountains. A sign for a dog-care service on the fault line advertised "canine coiffures and pawdicures," "dog day afternoon spa program," and "muttramony and barkmitzvahs."

The bizarre and mundane were joined here. On this day, more than one hundred feet underneath the park, they were digging in the damp darkness.

Just beyond the park at the northwest corner of Franklin and Camino Palmero was Al Jolson's mansion. Across the street and two doors up the hill was the Ozzie and Harriet Nelson residence, which was used for the exterior shots in the 1950s television show. One of the fault's strands passes beneath the living room.

Sandwich boards advertising "Star Maps" marked the beginning of Sunset Strip, where some very fancy establishments, such as Spago and the Chateau Marmont, were located in the fault zone. Sight-seeing buses with tourists glued to smoked-glass windows glided along the Strip, oblivious of the deeper realities of life on the edge.

At Havenhurst and Sunset I entered Dudley Do-Right's Emporium, a shop specializing in the TV-cartoon characters created by Jay Ward, such as Bullwinkle, Sherman and Peabody, Dudley, and Boris and Natasha.

I explained the seismic situation to the carefully dressed woman behind the counter, who seemed quite composed, on the surface. Her facial expression barely changed as she poured out a litany of continuing inner terror after the Northridge quake. But there was no place else to go that was warm, she said; certainly not San Francisco, where she had relatives.

At Sunset and La Cienega Boulevards, which is the center of The Strip, a borehole and rocks indicated that "the main strand of the Hollywood Fault lies either directly beneath or just south of Sunset Boulevard," according to

Dolan's paper, which was published in the *Geological Society of America Bulletin*. To the west, the fault splits first into two and then three strands as it enters Beverly Hills.

Lined up like expensive toy soldiers that might collapse in a row, should there be an earthquake along the faultline in this tony enclave, were structures of Mediterranean, French château, English baronial, New England colonial, and Southern plantation derivations.

Near the Beverly Hills Hotel at Rodeo Drive and Sunset, I could find only a slight swale across the wide street and carefully tended lawns that marked the western end of the fault. In Dolan's technical words, "There the pronounced south-facing scarps terminate."

UNIVERSAL CITY AND THE SUBCULTURE OF EARTHQUAKES

Over the years California has developed a culture of disaster, of which earthquakes are the dominant subculture. From the Hollywood Fault, I went looking for the cultural parallels to seismic events in the theme park at the opposite end of the subway tunnel, in films, and in literature.

My sampling began at Universal Studios Hollywood, overlooking the San Fernando Valley from the north side of the Santa Monica Mountains. It was here that people consciously exposed themselves to a tectonic thrill.

I called the studio and was connected to a publicity person. She told me to fax my request for information on the earthquake portion of the back-lot tram ride. I did. She telephoned the next day and said her boss wanted nothing to do with my endeavor. "We prefer not to be involved in a project that has something to do with real earthquakes," she said. "This is a theme park."

Of course, I thought, how perfect. We all prefer not to deal with the reality of disasters; and if we have to, best that they be presented as entertainment. Reality, however, has a way of intruding, even in a theme park. Universal Studios closed for a few days following the 1994 quake. One of the first places where the problem with steel welds was discovered was in a nine-story office building under construction in the entertainment complex.

I took the tram ride on a Sunday afternoon. Our upbeat guide was named John. I sat in the middle of a large group of Japanese tourists armed with cameras of every dimension. Why were they here? Surely, I thought, they were knowledgeable about real earthquakes. After all, Kōbe was only two years ago. Their capital city, Tokyo, had been destroyed by a quake and fire.

A number of disasters were simulated during the tram ride. None of them took place in southern California. Urban violence featuring King Kong was situated in New York City, a flash flood was located in the desert Southwest, a volcano erupted in the Northwest, and the location for the earthquake was San Francisco.

As we approached Sound Stage 50, John said:

> This is, in fact, the most unique sound stage in Hollywood. Inside this sound stage we have built the only two-level set in the world. It is being used now for a brand-new movie coming out next year. It's going to be called *Subway, Bloody Subway*. It's about a subway disaster that happens in the Bay Area.

We moved forward into an empty subway station and stopped. The tram began to shake.

> Oh, oh. One of those little California tremors. Hold on, hold on, folks. Something is coming. Look out for that truck. Oh no.

There were screams as the asphalt roadway above us unexpectedly collapsed, and a truck carrying liquid propane slid down the pavement toward the tram. Flames erupted. A few tourists managed to raise their cameras.

> Oh, my goodness, no. That train is coming into the station but the track is blocked. Put on the brakes!

A subway train collided with the tangled wreckage and buckled. There were more screams.

> Oh no! Look out! Don't be alarmed folks, it's not real water. It's just an illusion.

Sixty thousand gallons of recycled water gushed into the opposite side of the station from the conflagration. We pulled away from our constantly shaking perch just as the set began to retract for the next time.

"That was 'Earthquake, the Big One,'" said John, his voice deepening and becoming more menacing as he spoke the last three words.

Not wanting to leave us on that gloomy note, he added in a rising tone: "It was 8.3 on the Richter scale but 10 on the scale of fun, huh, guys?"

This was the year that disaster movies—whether of twisters, asteroids, volcanoes, or fires—were a Hollywood fad. *The New York Times* noted: "The point is, we love a well-contained disaster, and Hollywood is indulging this primal lust with a deluge of more than a dozen disaster pictures this year."

There had been such films in the past; three came to mind.

San Francisco, starring Clark Gable and Jeanette MacDonald, followed the Long Beach quake by three years. The closing earthquake scene was considered a classic. The screenplay by Anita Loos has been preserved in book form. The movie survives on late-night television and a videocassette.

The major studios were in their heyday in 1936 and had attracted top writers like Loos and F. Scott Fitzgerald. The writers liked the money but didn't relish this place where things went bump in the middle of the night. Loos grew up in San Francisco, was homesick for the Bay Area, and disliked Los Angeles—a fact she made abundantly clear in an early scene, when a drunken lout is identified as being a resident of that southern city.

The story concerned the salvation of an evil man, in this case a San Francisco gambler named Blackie (Clark Gable). The temblor was used as a redemptive device and was of secondary importance—the love story being the primary interest. Loos said, referring to her collaborator, "As an added feature we included the Great Earthquake, which, with true Frisco loyalty, Hoppy and I termed a 'fire.'"

The closing earthquake scene is a series of montages: music hall, city hall, restaurant, office building, business section, poor section. The Loos screenplay gave explicit directions: "During the above montage we see quick flashes in CLOSE SHOTS of terrified people in action indicated. Also, during the montage, the morning light gradually increases." Reality and myth are mixed. Brick buildings fall; a huge chasm opens in the street and swallows people.

Loos said the legendary producer, Irving Thalberg, insisted that every movie be a love story. So Blackie passes through the hell of an earthquake looking for Mary (MacDonald). In the process he "loses some of his vicious importunity" and becomes a better person. Blackie is able to pray at last.

Mary sees him praying. All ends well. Blackie gets Mary, and a new city—the 1936 version—is superimposed over the smoking ashes of the old one.

In the 1970s, it was Universal's *Earthquake,* starring Charlton Heston and Ava Gardner. The promotion copy on the back of that videocassette reads: "When the most catastrophic earthquake of all time rips through Southern California, it levels Los Angeles and sends shockwaves through the lives of all who live there."

The special effects created on the back lot of Universal Studios were quite convincing, for the most part, in the precomputer era. There is intense shaking, fire, and a flood when a concrete dam breaks. Los Angeles is a smoking and gutted city—a shell of its former self. The Charlton Heston character, an engineer, is heard to mutter through clenched teeth: "First time in my life I am ashamed of my profession. We should have never put up those forty-story monstrosities, not here."

An earthquake was the device used in 1996 to create a dystopian city-island along the derivative lines of *Blade Runner.* Through aftershocks, acid rain, thunder and lightning, and the rubble of Hollywood and Sunset Boulevards a two-gun, one-eyed Snake Plisskin ("Call me Snake," he hisses) struts in John Carpenter's *Escape from L.A.*

The movie begins with a presidential candidate predicting a millennium earthquake. A temblor measuring an improbable 9.6 on the Richter scale leaves the sinful city a smoking garbage dump shortly after noon on an August weekday in the year 2000. Familiar downtown buildings are made to bend and break, and freeways come crashing down on stalled vehicles with the help of computer graphics. Separated from the mainland by the shark-infested San Fernando Sea, Los Angeles becomes "an island of the damned." As Snake makes his way from the mainland to Los Angeles in a one-man nuclear submarine he speeds past a submerged Universal Studios.

Carpenter, who directed, wrote the screenplay, and composed the music, said he was influenced by the 1994 Northridge quake. I knew he was no stranger to natural disasters. He used to be my next-door neighbor on the Pacific Plate. First my house and then his burned to the ground in separate fires on the ridge adjacent to the San Andreas Fault.

The use of earthquakes in literature dates back to the Bible, where they were used to trumpet major events, such as the crucifixion, or served as metaphors. A sickly Robert Louis Stevenson, who lived in California for a time, wrote of life and death and earthquakes in the essay "Aes Triplex":

> We live the time that a match flickers; we pop the cork of a ginger-beer bottle, and the earthquake swallows us on the instant. Is it not odd, is it not incongruous, is it not, in the highest sense of human speech, incredible that we should think so highly of the ginger-beer and regard so little the devouring earthquake?

The 1906 San Francisco quake loosened a torrent of words, mostly journalistic. Earthquakes have been used mainly as a literary or structural device in fiction.

L. Frank Baum used a tornado to whisk Dorothy and Toto off in his first Oz book, but by 1908 and the third book in the series the San Francisco earthquake had captured the public's imagination. *Dorothy and the Wizard in Oz* begins with an earthquake. A boy named Zeb and a horse named Jim meet Dorothy at a railroad station. They proceed in a buggy toward a ranch.

Baum wrote:

> Neither the boy nor the girl spoke again for some minutes. There was a breath of danger in the air, and every few moments the earth would shake violently. Jim's ears were standing erect upon his head and every muscle of his big body was tense as he trotted toward home. He was not going very fast, but on his flanks specks of foam began to appear and at times he would tremble like a leaf.
>
> The sky had grown darker again and the wind made queer sobbing sounds as it swept over the valley.
>
> Suddenly there was a rending, tearing sound, and the earth split into another great crack just beneath the spot where the horse was standing. With a wild neigh of terror the animal fell boldly into the pit, drawing the buggy and its occupants after him.

They fall into dark space and land in a glass city, where Dorothy is reunited with the Wizard. The threat of earthquakes did not deter Baum; he moved to California in 1910.

Looking back in midcentury, Carey McWilliams commented:

How deeply the experience of living in an earthquake country has impressed the residents of the region is clearly shown in the novels that have been written about California. In many of these novels, one will find that the climax of the tale invariably is reached at precisely the moment when the dishes begin to rattle, the stove to bounce, and the chairs to dance.

In F. Scott Fitzgerald's novel *The Last Tycoon* an earthquake shakes a Hollywood studio and serves to bring the major characters together. Fitzgerald wrote in his proposal for the novel:

> He [an Irving Thalberg–type character named Monroe Stahr] has had everything in life except the privilege of giving himself unselfishly to another human being. This he finds on the night of a semi-serious earthquake (like in 1935) a few days after the opening of the story.

Fitzgerald may have meant the 1933 Long Beach temblor, as no earthquake of consequence was recorded in 1935. Except for the small hotels floating out to sea, Fitzgerald's description of an earthquake in the incomplete novel was authoritative:

> We didn't get the full shock like at Long Beach where the upper stories of shops were spewed into the streets and small hotels drifted out to sea—but for a full minute our bowels were one with the bowels of the earth—like some nightmare attempt to attach our navel cords again and jerk us back to the womb of creation.

Books that probed the darker side of the California experience began to appear in the late 1960s; but it was not until James D. Houston's novel *Continental Drift* was published in 1978 that one was structured around tectonic motion. Houston wrote of his protagonist, Montrose Doyle, who lives on a ranch near Santa Cruz that is bisected by the San Andreas Fault:

> Most of the time he doesn't think about it at all. It is simply there, a presence beneath his land. If it ever comes to mind during his waking hours, he thinks of it as just that, a presence, a force, you might even say a certainty, the one thing he knows he can count on—this relentless

grinding of two great slabs which have been butting head-on now for millennia and are not about to relax.

Earthquakes also could serve as a source of humor. The main character in a serial novel, *Ear to the Ground*, that ran in the *Los Angeles Reader* in 1995 is Charlie Richter, the fictional grandson of the famed Caltech seismologist. Its authors, Paul Kolsby and David L. Ulin, wrote:

> Charlie's new employer, the Center for Earthquake Studies, or CES, was endowed with a multimillion-dollar budget rumored to have come about, in part, through a hushed yet symbiotic relationship with the entertainment industry, whose interest lay in the Earthquake Channel, as well as an interactive TV series called *Rumble*. "If the Big One hits L.A.," mused an inside source, "the studios will be in on the ground floor."

FROM SAN BERNARDINO TO THE GULF OF CALIFORNIA

Following my excursions through the back lots of Hollywood, I set off on the last leg of my journey on the San Andreas rift. It was 115 miles from Cajon Pass to the Salton Sea, where the fault ends. That termination, however, lacks completeness; so I decided to travel the extra distance across the desolate Colorado River delta in Mexico to where the plaited braids of the fault system dip into the Gulf of California.

I drove east from Los Angeles during the week a local newspaper announced that three-quarters of the homes damaged in the Northridge quake had been repaired. This event was represented as a victory. A local television station ran what-are-they-doing-now-type interviews with victims of the three-year-old quake. The familiar scenes of carnage were shown as promos. It seemed as if the station yearned for those dramatic days and nights—and the higher ratings.

Also that week, a low-level volcano and earthquake alert was issued for the Mammoth Lakes region, which served southern Californians as a recreation resort on the east side of the Sierra Nevada.

Two hundred eighty-four small earthquakes were detected at Caltech that week. Two were noticeable. One was felt in Palm Springs; the other rattled dishes on the border. All in all it seemed like an average week, seismically speaking.

I arrived in San Bernardino on an overcast Saturday afternoon in February. Like the San Francisco Bay area, San Bernardino is bisected by two major faults—the San Jacinto and the San Andreas, the former being a branch of the latter. The Inland Empire, as the region styles itself, was assigned the highest probability for a major earthquake on a recognized fault in southern California.

There was a slight drizzle, but not enough to deter the launching of scores of colorful hang and paragliders from the mountain heights behind the California State University campus. They carved rainbow-hued arcs in the air as they descended toward Andy Jackson Air Park, a custom-designed landing zone for such craft located directly on the fault line.

I could see the fanciful airframes as I took Northpark Boulevard past the university and turned right on the first road after the dike. I headed toward the elevated ramp that marked the landing zone. The treeless terrain was scarred by dirt bikes. The landscape was also flood and fire prone. There was a standing order for gliders to land immediately if fire-fighting aircraft appeared in the area.

With the relatively flat rift zone below, high mountains above, prevailing southwest winds, and a strong sun to heat the air and form thermals, it was possible to glide for five hours and rove for forty miles along the front of the range.

"We all know we are on the fault line," said Rob Von Zabern, president of the Crestline Soaring Society. That did not deter the many winter visitors from Europe and elsewhere who flocked to one of the best soaring sites in the world, which had been formed by the convulsions of the earth.

Ann Strawn and her husband, Glenn Drewes, also knew without a doubt that they were on the fault line. However, they did not make that discovery until after they had purchased their home in Highland, a hillside suburb of San Bernardino. They had been living directly in the middle of the rift zone for about one year before I found them gardening in their spacious backyard.

Strawn explained, "A friend of my husband visited and said, 'You know, this is on the fault.' We blinked twice and said, 'Oh, I bet you're right. That's why it was available.' "

So much for the legal requirement of disclosing the presence of the fault to home buyers, I thought. A study by the relevant state agency, I later read, deemed such disclosure requirements to be ineffective.

I asked how they felt about living directly on the fault line. "Well, my husband doesn't care, but I get a little queasy sometimes," Strawn said. They were both native southern Californians and used to tremors. As a child growing up in San Diego County in the early 1960s, Strawn had seen an earthquake advance across the landscape as a visible wave.

The computer programmer and the biology teacher had been looking for an older home when they found the stucco-and-tile-roof residence, built in 1969. That was three years before passage of a state law that now requires structures used for human occupancy to be set back a minimum fifty feet from an active fault.

There were, however, subtle clues to the presence of the fissure. A sign across the street noted the existence of Rancho San Andreas, a subdivision. The water table was extremely high. None of the newer homes in the neighborhood trespassed on the U-shaped depression that allowed the couple to plant a garden in relative isolation from their neighbors.

I pointed out that this section of the fault had been quiet since 1857. Strawn correctly deduced, "So it's due." They did not carry earthquake insurance. "Between the cost and the deductible it's ridiculous," she said.

The surrounding new cheek-by-jowl homes, through which the green swath of the fault zone passed, were named to fit marketing needs, not geologic realities: Barcelona, Cottages of Corsica, Parkcrest Series, Centrex Homes, Club View Villas, Lakecrest Estates. The broad streets were alive with flapping subdivision pennants; but the neighborhoods were strangely vacant on a weekend afternoon. "Protected by" signs, indicating the preferred security service, proliferated on green lawns.

At Church Street and Oak Ridge Road I came across one of the clearest expressions of the San Andreas Fault that I had seen in all my travels. It was also a classic California scene. Land was about to change uses and increase in monetary value while food was taken out of production.

Beyond Church Street there was a neglected orange grove. The fruit was rotting on the trees, and rusted orchard heaters lay on the ground. The grove was the next candidate for a subdivision unit that undoubtedly would be given an exotic name.

Bisecting one end of the orange grove on an east-west axis and extending to the west of Church Street toward the couple's property. was a one-hundred-foot-wide trench with a berm on the downslope side. It cut a wide path through the backyards of the newer homes. They were fairly expensive structures, this being the heights, where there was a view on a rare smog-less day.

I wondered if the owners had been told about their tectonic neighbor; I didn't think so. I sometimes felt like the bearer of bad news, or, at the very least, one who confirmed it. Knowing the fate of the messenger, this time I chose not to knock on any doors.

To the east the fault crossed under both secular and religious institutions. At the Forest Service's Mill Creek Ranger Station on Highway 38 the receptionist said, "Rumor has it that the fault goes right under here." She looked up at the heavy beams and electrical fixtures and added, "It would not be great if there was an earthquake." Farther to the east the fault passed through the Oak Glen Christian Conference Center, where the signs declared, "Thou shalt not park here."

At the old Oak Glen schoolhouse, old in southern California meaning built in 1927, the back of the cobblestone structure had collapsed in the Landers quake. It had since been repaired. Lucille Wilshire Broaders, whose grandfather homesteaded in the area, said earthquakes altered the flow of the springs. Of the 1992 temblor, she said, "There was dust and rocks and roaring. I went out and all I could see was the mountains full of every-thing. I could see new houses cracked and damaged even before they were occupied."

Farther east the freeway climbed gradually to San Gorgonio Pass. North of Interstate 10 the faulting became complex. It was here that the Big Bend section ended and the fault system began to take a more southerly course. After citing the appropriate study, a geological tour guide to the area declared that "San Gorgonio Pass constitutes a structural knot in the mod-ern San Andreas Fault zone." The pass also marked the dividing point between the coastal basin and the desert.

I stopped in nearby Palm Springs to do some research in the well-appointed public library. Signs on the ends of the book stacks noted: "Warning: During earthquakes leave stack areas and take cover under tables."

I could understand why. Libraries had not fared well in the Northridge quake. A fully loaded stack of books could weigh close to one ton. A study of the 1994 event determined: "A mere nine such rows set into motion during an earthquake represent more than fifty tons of weight transferring its energy to the building. Large libraries may have hundreds of such rows of book stacks."

I moved quickly from the stacks after making my selections, for the irony of being caught unaware beneath a load of books about earthquakes did not escape me.

The librarian in charge of clipping newspapers said she did not bother to save stories about the smaller shocks. There was, however, a fair amount of material on the North Palm Springs quake of 1986, as well as the nearby Landers temblor.

I found a story about the public relations director of one of the Palm Springs luxury hotels who was working with a geologist to package fault-line tours for Japanese tourists after the Landers quake. The article ended: "She said that reaching out to groups interested in earthquakes is one way of making the best of a shaky situation." What was there that encouraged punning about earthquakes? I wondered.

I drove across the Coachella Valley to Desert Hot Springs, through which the north branch of the fault passes, and took Dillon Road east past one spa, motel, mobile-home park, and recreational-vehicle resort after another—all offering the warm, healing waters provided by the fault line.

The prevalence of water in the rift zone is most noticeable in the desert. This life force and the types of vegetation that are dependent upon it are relatively abundant all along the fissure. Either a fault block drops and traps the water, or the impervious layer of one side of the fault dams the flow of underground water. Pressure forces it to the surface along fracture lines.

The two branches of the fault—designated Mission Creek to the north and Banning to the south—pass on either side of the Indio Hills. On both sides there are palm tree oases. These junglelike grottoes in the middle of the desert resemble green beads strung on a seismic thread.

The largest patch of greenery was Thousand Palms Oasis, reachable off the road of the same name. The oasis, now part of the twenty-one-thousand-acre Coachella Valley Preserve, was purchased in 1984 by the

Nature Conservancy. Since then seismologists have dug trenches and discovered evidence for two prehistoric earthquakes—a "penultimate" event sometime between 1040 and 1410 and a smaller disturbance around 1680. Since then this segment of the fault has been relatively quiet.

The bedrock on one side of the fault formed an underground dam, trapping water flowing from higher elevations. The thick groves of native *Washingtonia filfera* fan palms were dependent on the water that was forced to the surface, since their roots seldom penetrated deeper than twelve feet. Along with palms there were willows, cottonwoods, mesquite, and salt grass.

The ponds provided a refuge for the endangered desert pupfish, who have been preyed upon elsewhere in the desert by nonnative species. Mammal and bird life was intense; shade and water were provided for the weary desert traveler.

Driving south on Interstate 10, I encountered a fair amount of freeway construction centering around interchanges. Orange Caltrans signs with pictures of idealized seismograms and bridges signified earthquake retrofitting projects. The fault crossed the freeway between the Dillon Road interchange and the Coachella Canal, at approximately the point where a sign designated the elevation as being sea level.

South of the freeway the best place to view the fault line is at the mouth of Painted Canyon in the Mecca Hills. At the town of Mecca, which had a decided Mexican flavor, I drove east on Sixty-sixth Avenue past the Riverside county dump and Coachella Canal to Painted Canyon Road, the first left after the canal. The graded dirt road led to the secluded canyon.

Here in a spectacular badlands setting the San Andreas Fault was distinguished by reddish brown clay. The band of fault gouge was 150 feet wide. It was late afternoon on a cloudy winter day. For a brief moment the sun pierced the thick clouds above the peaks of the distant San Jacinto Range, illuminating the crenelated hills that glowed flame red like a besieged medieval city. Three jets roared overhead on their way to a nearby bombing range.

To get a feeling for the end of the fault, I camped beside the Salton Sea for a few days. I don't recommend it. The 360-square-mile receptacle for the agricultural waste water from the Imperial Valley is highly saline. The beach and offshore waters were thick with dead fish, on which thousands of gulls feasted. Car and truck traffic was heavy on nearby Highway 111. There were numerous freight trains. Both modes of transportation straddled the adjacent fault line.

From the Salt Creek Campground I walked east a couple of hundred yards into the desert until I hit the creek bed and then followed it to where a side canyon headed south. There was one lone palm tree and some vegetation in the creek. The detailed USGS map I was using noted "lithologic contrast" and "small collapse fissures on trace" along the side canyon.

I returned and recrossed the highway at the point where I had previously noticed two highway patrolmen working a speed trap in separate cars. First one ticketed speeding Saturday-morning motorists while the other acted as a backup. Then they reversed chores.

I walked up to the parked white patrol car in which one of the officers sat. It was marked "Highway Patrol K-9 Unit." I looked into the back of the vehicle but could not see past the smoked glass.

I asked if he had a dog back there, and the patrolman answered: "Yup. Best partner I ever had." He was reading a manual on police K-9 dogs.

"I'm looking for the San Andreas Fault," I said. "Could you direct me to it?"

He said it was over on the other side of the mountains.

Wrong, I said to myself, hoping that he had a better command of canine needs.

That evening I drove south a few miles to Bombay Beach and ate at the Ski Inn. The fault bisected the small retirement community and disappeared from maps at that point. No one was very worried about earthquakes. Like the Dutch, they gave more thought to the dike that kept the rising waters of the Salton Sea from further inundating their homes. Of course, a quake could easily destroy the dike.

Thursday night was tostada night at the Ski Inn, and snowbirds from surrounding trailer parks flocked to the smoky bar. I sat between three brothers, one of whom owned the bar. He asked, "Do you see anyone here who seems to be afraid of an earthquake?" I looked around and allowed that I didn't.

That night the Hale-Bopp comet was the dominant presence in the translucent desert sky. I followed its progress from my sleeping bag and, recalling the year of 1812, wondered if it was a harbinger.

Early the next morning I drove through the rich farmlands of the Imperial Valley, which had experienced its share of quakes, and into Sonora, Mexico, where I picked up the end of the San Andreas Fault system in the sere Colorado River delta. There is no clear matchup between the end of the San Andreas Fault at the Salton Sea and other faults in the valley and Sonoran Desert. However, the wider system of intertwined faults that make up the tectonic boundary between the two plates passes through the Salton Trough.

The moonscape of the delta was familiar terrain to me. Pursuing various writing projects and my own curiosity over the years, I had kayaked the remnants of the river and the head of the gulf, flown over the flat deltaic landscape in a small plane, and hiked across its mirage-filled surface to a deserted nineteenth-century shipyard. I find the delta the most forbidding and fascinating corner of the North American continent.

South of Riito, on the road to El Golfo de Santa Clara, I thought about taking the maintenance road bordering the concrete-lined Wellton-Mohawk bypass drain to Santa Clara Slough. I decided to stick to the paved road because time was short, I didn't know the condition of the dirt road, and I suspected that the tall cattails at its end would limit views. Instead, I stopped at different points on the bluff that overlooked the slough.

The Cerro Prieto Fault formed the bluff, known as the Gran Desierto Escarpment. Numerous small springs surrounded by patches of grass and grazing livestock marked its passage along the foot of the slope. The sources of water for the slough, the largest wetland in the Sonoran Desert, are extremely high tides in the gulf and agricultural waste water from southern Arizona. The drain was constructed by the United States in order to improve the quality of water being delivered to agricultural fields in Mexico from the Colorado River.

There are accounts of earthquakes along the Cerro Prieto Fault as far back as November 29, 1852, when a column of steam more than one thousand feet in height shot from the Volcano Lake area. It was spotted forty-five miles to the northeast at Fort Yuma, where an army lieutenant reported: "Large openings were made in the ground all around us, and water and steam thrown up in large quantities."

On the day in 1857 when the mighty Fort Tejon temblor struck far to the northwest in central California a steamboat was anchored at the mouth of the Colorado River. One of its crew wrote home:

> I must tell you about the earthquake we had down the river. The boat rocked so that we could hardly stand. We looked up the river and the water all drawed off of one place and left it dry. Then in a moment all rushed back again foaming and tumbling.

The delta is an appropriate setting for strange phenomena. Given the right conditions, a wall of water known as a tidal bore rushes up the meandering channel. That night I again saw the comet and decided that it was time to head home toward my end of the San Andreas Fault.

A few weeks later I took a long-neck bottle of Mexican beer out onto the deck to enjoy the tranquil view of the San Andreas Fault. I thought about my journey through seismic history and the land of natural disasters.

The current situation did not bode well. There was disarray in the seismological community along with some indications of change. Much proceeded as usual as the millennium approached.

The National Research Council had formed a Committee on the Science of Earthquakes to determine priorities. Thomas H. Jordan, a geophysicist at the Massachusetts Institute of Technology, was the chairman. In an editorial in a seismological journal, Jordan said: "The collapse of earthquake prediction as a unifying theme and driving force behind earthquake science has caused a deep crisis."

There really wasn't much that I could do except take another swig of beer, check where all the shut-off valves were located, and secure breakable items. I also should make some inquiries about the neighborhood disaster plan.

In the end, it came down to a case of personal responsibility and good fortune. I fervently hoped that I would miss the inevitable moment when California finally ran out of luck.

APPENDIX

THE RICHTER SCALE IS NO MORE

One of the minor goals I had in mind while writing this book was to see if I could complete the text without mentioning magnitudes for specific seismic events. Except for references to a generic magnitude 8 earthquake, I have accomplished this objective and believe the book does not suffer for lack of these mysterious numbers.

If there is one reason why laypersons like myself cannot comprehend the raw power of earthquakes, cannot make a meaningful comparison between quakes, and cannot relate the energy of a given temblor to, say, a tornado, it is because of the awkwardness and incomprehensibility of this arcane system of measurement.

The concept of magnitude is a good example of the inability of the vast majority of seismologists to communicate adequately with the general public. In a study of seismology and communication, Arnold Meltsner of the University of California at Berkeley wrote of the seismological establishment: "They are not, however, well equipped for communication with a wider public. . . . The problem of how to convert or translate scientific language into ordinary language is fundamental."

I have read and puzzled over countless explanations of the system for measuring earthquakes. Learned people have attempted to explain the magnitude scale to me in person and on the Internet. I got into a heated argument with one seismologist who claimed that when he gave public talks the audience understood the concept. I didn't believe him but still felt dumb.

Then I read that John McPhee, who has written extensively on matters of geology, had "no idea how the scale works." He wrote in *Assembling California*, "Richter was a professor at Caltech. His scale, devised in the nineteen thirties, is understood by professors at Caltech and a percentage of the rest of the population too small to be expressed as a number." I felt better.

The Richter scale came about in the following manner. Beno Gutenberg and his younger colleague, Charles F. Richter, borrowed the concept of magnitude from astronomers and applied it to earthquakes. At the time, Richter wrote there was no scale "to discriminate between large and small shocks on a basis more objective than personal judgment."

A young science that was striving for acceptance quickly embraced the concept, which lent it the air of objective quantification. By the 1950s, the Richter scale was being widely used in press reports of earthquakes. "Lately there have been complaints that the use of the magnitude scale is confusing, or at least that the reporting of magnitude in the newspapers 'confuses the public,' " Richter said.

While a reporter on the *Los Angeles Times*, I called Professor Richter a couple of times at his office in Pasadena to obtain the magical numbers for minor quakes. To talk with *the* Dr. Richter was a bit daunting for a young reporter with little scientific background. I don't remember what he said. I do recall that I parroted exactly what he told me.

The Richter scale is now obsolete, but for fifty years it served its purpose and its name became synonymous with the measurement of earthquakes. There are now four scales with numerous variations. The moment magnitude scale has gained widespread acceptance in the scientific community for the purposes of reporting numbers to the public. The media frequently attach the name Richter to this scale out of laziness and habit.

The clearest explanation of modern magnitudes that I have been able to locate is by Lucile M. Jones of the Pasadena office of the USGS in the publication *Putting Down Roots in Earthquake Country*. It almost makes sense to me. She wrote:

More recently, seismologists have shown that magnitude is proportional to the energy released in the earthquake. A magnitude 6.0 earthquake has about 32 times more energy than a magnitude 5.0 and almost 1,000 times more energy than a magnitude 4.0 earthquake.

Seismologists measure different earthquake "dimensions" with different magnitude scales. Each scale measures how fast the ground moves at a different distance and in a different frequency band. Each has its uses, but all are limited because they measure only part of the ground motion.

In recent years, seismologists have developed a new scale, called moment magnitude, that avoids many of these limitations. Moment is a physical quantity related to the total energy released in the earthquake. It can be estimated by geologists examining the geometry of a fault in the field or by seismologists analyzing a seismogram. Because the units of moment are cumbersome, it has been converted to the more familiar magnitude scale for communication to the public.

But imprecision is rife within a given scale. The moment magnitude for a given quake can vary slightly, depending on who reads the seismogram and what formula is applied.

Moment magnitudes for earthquakes mentioned in the text follow in chronological order. Question marks indicate approximations:

Lisbon, Portugal, 1755	8.7?
Santa Ana River, Calif., 1769	6.0?
New Madrid, Mo., Dec. 16, 1811	8.1?
New Madrid, Mo., Jan. 23, 1812	7.8?
New Madrid, Mo., Feb. 7, 1812	8.0?
San Juan Capistrano, Calif., 1812	7.0?
Santa Barbara, Calif., 1812	7.0?
Fort Tejon, Calif., 1857	7.8
Owens Valley, Calif., 1872	7.8?
San Francisco, Calif., 1906	7.7
Long Beach, Calif., 1933	6.4
Kern County, Calif., 1952	7.5
Chile, 1960	9.5
Anchorage, Alaska, 1964	9.2

San Fernando, Calif., 1971	6.7
Haicheng, China, 1975	7.0
Tangshan, China, 1976	7.5
Whittier Narrows, Calif., 1987	5.9
Loma Prieta, Calif., 1989	7.0
Landers, Calif., 1992	7.3
Northridge, Calif., 1994	6.7
Kōbe, Japan, 1995	6.9

To the uninitiated, like myself, these numbers are deceptively slight, relate poorly to each other, and rob earthquakes of their awesome power. After the 1989 Loma Prieta earthquake, Arch C. Johnston, a seismologist who specializes in the New Madrid seismic zone, wrote in a USGS publication:

> Let's face it: seismologists do a pretty poor job of communicating the facts about our science to the public. Earthquake magnitude is the classic example. How many of us have struggled to explain the Richter scale? We explain that it is logarithmic, with each unit indicating a factor of 10 increase, but this really represents a factor of 32 increase in intrinsic earthquake size, and in any case we don't use the Richter scale anymore. By then the unfortunate listener is reeling and can be dispatched quietly by mentioning negative magnitudes or saturation. We even wonder why the audience or the reporter has this glazed look when we finish.

Johnston constructed a linear magnitude scale that gave a good indication of the relative strengths of different seismic events. Given the moment magnitudes of four earthquakes, here are their equivalents on Johnston's Earthquake Strength Scale:

Threshold of damage, 5.0	1
Loma Prieta, 1989, 7.0	100
Alaska, 1964, 9.2	10,000
Chile, 1960, 9.5	31,600

In order to compare energy released from earthquakes to other types of natural disasters, Johnston assigned such events moment magnitudes. They are: a tornado, 4.7; Mount Saint Helens volcanic eruption, 7.8; a well-developed hurricane over a ten-day life span, 9.6.

ACKNOWLEDGMENTS

This book was the idea of Carl Brandt, the literary agent who represents me. I had finished a book on the natural and human history of California titled *The Seven States of California*. It contained a chapter on the central portion of the San Andreas Fault.

Two other book ideas of mine had failed to fly. Carl then suggested a book on the fault and earthquakes. It seemed like a natural for this rift-line resident and veteran of a number of California disasters. I should add that Carl is more than an agent; he is a buffer, sounding board, and career guidance counselor.

A number of other persons have helped me with various book projects over the years. My wife, Dianne, read the manuscript with her usual acuity and again provided the income that allows me to write. My good friend Doris Ober, a freelance editor, is my first professional reader. She supplies overall guidance with a great deal of insight, kindness, and wit. This time she was part of the story. Then there are the other readers, all intelligent friends and fellow San Andreans who have supplied insightful comments. They are Richard B. Lyttle and Connie and Michael Mery.

At Henry Holt, Jack Macrae's intuitive grasp of what was needed to improve the text has resulted in a better book. We both discovered the use and limits of E-mail for communicating ideas. The copyeditor, Adam Goldberger, caught many silly and some stupid mistakes. Michelle McMillian fashioned a superior design.

At the start of this project a former colleague on the *Los Angeles Times*, Kenneth Reich, gave me an assessment of who in the seismological community might be helpful. His printed list of names and phone numbers was invaluable. Ken is the only daily journalist who specializes in earthquakes.

When I ran up against the fact that no one had collected the earthquake stories of Native Americans, Malcolm Margolin of Heyday Books in Berkeley, publisher of *News from Native California*, recruited some Cal students who researched that aspect for me. Gerald Haslam, James D. Houston, and Charles Wollenberg helped me with literature searches.

I strive for accuracy, but I was a stranger to the disciplines of seismology, geology, and earthquake engineering. So I asked the experts for help. A friend and former state geologist of Utah, Genevieve Atwood, read the entire manuscript. She caught a number of errors and differed strenuously with my criticism of the profession. Kerry Sieh and Thomas H. Heaton of Caltech, Lucile Jones and Robert Wallace of the USGS, and Tousson Toppozada of the California Division of Mines and Geology read portions of the text and commented on technical matters and interpretation. I listened carefully to what they all had to say; the final product, however, is my responsibility.

When I arrived at some tentative conclusions, I tried them out on the experts. They were Jones and James J. Mori of the Pasadena office of the USGS; Heaton, Clarence Allen, Hiroo Kanamori, George Housner, and Wilfred D. Iwan of Caltech; Wallace, David P. Schwartz, William Ellsworth, and James H. Dieterich of the Menlo Park office of the USGS; Gregory C. Beroza of Stanford University; and Lloyd S. Cluff of the California Seismic Safety Commission. Needless to say, there was agreement and disagreement; sometimes the latter was heated.

I spent a good deal of time in various libraries in California. The Earth Sciences Library at the University of California's Berkeley campus, where seismology got its start in this country, was the single best repository. I finished my research just as it moved into temporary quarters in July of 1997, when the second attempt to retrofit McCone Hall got under way. On the main campus, the Bancroft and Life Sciences Libraries were other frequent stops. Another excellent source was the Earthquake Engineering Research Center's library at the university's Richmond Field Station. Charles D. James, information systems manager of that library, went out of his way to help.

Elsewhere in the Bay Area, the Governor's Office of Emergency Services in Oakland has some interesting documents that should be available in a more central repository. In Menlo Park there was the USGS western region library and nearby was the Branner Earth Sciences Library on the Stanford University campus.

In southern California, the earth science library at Caltech was the single best source. Nearby was the Huntington Library. I also did some work on the past history of earthquakes in the Special Collections Section of UCLA's main library.

I joined the Internet for this project and became part of the on-line community. There are a host of Web sites devoted to earthquakes that are best accessed by a search engine. The main USGS Web site—www.quake.usgs.com—is oriented toward northern California; but it provides links to other sites that focus on seismic activity around the world.

Two news groups, ca.earthquakes and sci.geo.earthquakes, provided me with a wealth of information and served as focus groups when I posed a question or stated a tentative conclusion. The USGS posts weekly seismic reports for northern and southern California on the two sites. From the ravings of far-out pseudoscientists, to the patter of knowledgeable seismologists from around the world, to comments from just plain citizens, the two sites—particularly sci.geo.earthquakes—gave me a range of opinions that would not otherwise have been available.

These electronic research services will not replace hard copy, but they are a valuable adjunct. I thank the contributors. I have preserved their privacy, except where I received permission to use their comments.

I was given refuge in my travels. Elizabeth Zarlengo and Gary Ireland made their apartment in San Francisco available when they were not there so that I didn't have to return home every night when working to the south on the peninsula. In Los Angeles, my mother-in-law, Louise Caccioli, lent me her spare bedroom when I was in southern California. The condominium was damaged in the 1994 quake, and I have watched the neighborhood slowly recover over the years. It sits on the alluvial deposits of the adjacent Los Angeles River. But I slept well.

Elsewhere along the eight-hundred-mile fault line, except when I was home, I camped outdoors, as is my custom. This time I had no enclosed van, but rather the shell on the back of a pickup truck and a tent. I prefer to travel in this manner when researching a book or magazine article. It puts me in closer contact with my subject and gets me out of disinfectant-scented motel rooms and away from fat-laced café cooking. I had no unpleasant experiences, and for part of the time I had a comet for company at night.

SUGGESTED READING

For the lay reader who wants to delve further into the subject but would like to avoid tiresome technical details, I recommend the following: *Tales of the Earth* (Charles Officer and Jake Page) sets the scene. On the science of earthquakes there is the simplistic *Why the Earth Quakes* (Matthys Levy and Mario Salvadori) and the more complex *Earthquakes* (Bruce A. Bolt). *When the Snakes Awake* (Helmut Tributsch) will make you a believer in earthquake phenomena. For the 1906 San Francisco earthquake *Denial of Disaster* (Gladys Hansen and Emmet Condon) and the report of the state commission, *California Earthquake of April 18, 1906* (Andrew C. Lawson, the Carnegie Institution, Washington, D.C.), are first-rate. *Peace of Mind in Earthquake Country* (Peter I. Yanev) and *Earthquake Survival Manual* (Lael Morgan) deal with safety. *Putting Down Roots in Earthquake Country* (Lucile M. Jones, Southern California Earthquake Center, University of Southern California) is the best pamphlet. USGS map sales stores have a host of publications, some of which are free. Although not free and somewhat technical, *The San Andreas Fault System, California* (Robert Wallace) is a good reference work. To provide the proper ambiance for reading, purchase the audiotape *Earthquake Sounds* (Karl V. Steinbrugge, Seismological Society of America, Emeryville, Calif.).

For the fault-line tourist I recommend the following maps and guides: the regional auto club maps for roads, the northern and southern California atlases and gazetteers published by the DeLorme Mapping Company for topography and the fault zone, and for the precise location of the fault in relation to the roads and topography earlier out-of-print editions of *Earthquake Country* by Robert Iacopi. The new 1996 edition lacks the wonderful maps and aerial photos on which the fault line was traced in earlier versions.

SOURCE NOTES

What follows are the sources I consulted, listed under the chapter heading and sub-head section where the material appears. Some journal titles have been abbreviated: *Bulletin of the Seismological Society of America (BSSA)*; *Seismological Research Letters (SRL)*; *Journal of Geophysical Research (JGR)*; *Geological Society of America Bulletin (GSAB)*. No page numbers are cited if the work was generally relevant. Should more precise citations follow the first mention, they address specific matters.

1. SAN FRANCISCO (THE PRESENT), PAGES 3–8

The Day California Ran Out of Luck

USGS, Karl V. Steinbrugge et al., *Metropolitan San Francisco and Los Angeles Earthquake Loss Studies: 1980 Assessment*, Open-File Report 81-113, Washington, D.C., 1981; California Division of Mines and Geology, James F. Davis et al., *Earthquake Planning Scenario for a Magnitude 8.3 Earthquake on the San Andreas Fault in the San Francisco Bay Area*, Special Publication 61, Sacramento, Calif., 1982; Fouad Bendimerad et al., *What If the 1906 Earthquake Strikes Again? A San Francisco Bay Area Scenario*, Risk Management Solutions, Menlo Park, Calif., 1995; California Division of Mines and Geology, Karl V. Steinbrugge et al., *Earthquake Planning Scenario for a Magnitude 7.5 Earthquake on the Hayward Fault in the San Francisco Bay Area*, Special Publication 78, Sacramento, Calif., 1987; Tousson R. Toppozada et al., "Planning Scenario for a Major Earthquake on the Hayward Fault," in California Division of Mines and Geology, Glenn Borchardt, ed., *Proceedings of the Second Conference on Earthquake Hazards in the Eastern San Francisco Bay Area*, Special

Publication 113, Sacramento, Calif., 1992; USGS, Working Group on California Earthquake Probabilities, *Probabilities of Large Earthquakes in the San Francisco Bay Region*, USGS Circular 1053, Washington, D.C., 1990; Gladys Hansen and Emmet Condon, *Denial of Disaster: The Untold Story and Photographs of the San Francisco Earthquake and Fire of 1906*, Cameron and Company, 1989, pp. 135–153. Material on the San Andreas and Crystal Springs Dams is cited in the notes for the Loma Prieta chapter. The description of the dam failures was drawn from material on dams in the Fort Tejon, Loma Prieta, and Northridge chapters.

2. TERRA NON FIRMA, PAGES 9–16

Earthquake Country

USGS, Robert E. Wallace, ed., *The San Andreas Fault System, California*, USGS Professional Paper 1515, Washington, D.C., 1990; Robert E. Powell, *The San Andreas Fault System: Displacement, Palinspastic Reconstruction, and Geologic Evolution*, Memoir 178, Geological Society of America, 1993; Robert M. Norris and Robert W. Webb, *Geology of California*, John Wiley & Sons, 1990, pp. 436–458; Robert Iacopi, *Earthquake Country*, Lane Publishing Co., 1971, and Fisher Books, 1996; Thurston Clarke, *California Fault*, Ballantine, 1996; Peter Fish, "It's Our Fault," *Sunset*, June 1995; "The San Andreas Fault," Sandra S. Schulz and Robert E. Wallace, USGS, Washington, D.C., 1989; "Knocking New York," *New York Times*, May 17, 1995; Mason L. Hill, "San Andreas Fault: History of Concepts," *GSAB*, March 1981. Along with the map in the book edited by Wallace, 1990, there are two recent maps of the fault line: California Division of Geology and Mines, 1994, and National Geographic Society, 1995.

For this short explanation of plate tectonics I relied on H. E. LeGrand, *Drifting Continents and Shifting Theories*, Cambridge University Press, 1994; Eldridge M. Moores, ed., *Shaping the Earth*, W. H. Freeman, 1990; and Eldridge M. Moores, "The Story of Earth," *Earth*, December 1996.

The Nature of Seismic Events

For the science and basic characteristics of the earth, earthquakes, and their comparisons to other natural upheavals I relied on Shawna Vogel, *Naked Earth*, Dutton, 1995; Geof Brown et al., *Understanding the Earth*, Cambridge University Press, 1992; Edward A. Keller, *Active Tectonics*, Prentice-Hall, 1996; Bruce A. Bolt, *Earthquakes*, W. H. Freeman, 1993; Stephen L. Harris, *Agents of Chaos*, Mountain Press, 1990; Matthys Levy et al., *Why the Earth Quakes*, W. W. Norton, 1995; Gilbert F. White et al., *Assessment of Research on Natural Hazards*, MIT Press, 1975; Charles Officer and Jake Page, *Tales of the Earth*, Oxford University Press, 1993; Andrew

Robinson, *Earth Shock*, Thames and Hudson, 1994; Barbara Tufty, *1001 Questions Answered about Earthquakes, Avalanches and Other Natural Disasters*, Dover Publications, 1978; Robert L. Kovach, *Earth's Fury*, Prentice-Hall, 1995; E. L. Jones, *The European Miracle*, Cambridge University Press, 1981, pp. 22–41; David Alexander, *Natural Disasters*, UCL Press (London), 1993; and Robert S. Yeats et al., *The Geology of Earthquakes*, Oxford University Press, 1997.

The information on ancient earthquakes came from Helmut Tributsch, *When the Snakes Awake*, MIT Press, 1984, pp. 185–202; Derek Ager, *The New Catastrophism*, Cambridge University Press, 1995, p. 168; Officer, 1993, pp. 28–30, 33; Amos Nur, lecture: "Armageddon's Earthquakes," fall meeting, American Geophysical Union, 1996; Amos Nur, "Earthquakes and Archeology in the Eastern Mediterranean and Ancient Near East," Stanford University lecture, May 19, 1997; Dorothy B. Vitaliano, *Legends of the Earth*, Indiana University Press, 1973, pp. 88–91, 179, 232, 251–267; John Antonopoulos, "The Great Minoan Eruption of Thera Volcano and the Ensuing Tsunami in the Greek Archipelago," *Natural Hazards*, March 1992.

Casualty and damage figures were derived from Alexander, 1993, pp. 5, 464–465; Jones, 1981, p. 24; Robinson, 1994, pp. 7–8, 48, 78; USGS, Carl W. Stover and Jerry L. Coffman, eds., *Seismicity of the United States 1568–1989*, rev. ed., USGS Professional Paper 1527, Washington, D.C., 1993, pp. 113–181; Bolt, 1993, pp. 271–273. Risk came from Robinson, 1994, p. 9, and "Ten Risky Places," an extract from Mark Monmonier's *Cartographies of Danger*, University of Chicago Press Web site, 1997.

Early History and Mythology

For the early history and mythology of earthquakes I consulted L. Don Leet, *Causes of Catastrophe*, McGraw-Hill, 1948, pp. 11–24, 43–47; Maria Leach, ed., *Funk and Wagnalls Standard Dictionary of Folklore, Mythology and Legend*, Funk and Wagnalls, 1972, p. 334; Vitaliano, 1973, pp. 81–86; Tributsch, 1984, pp. 59–60, 133, 140, 174–175; Frederic Golden, *The Trembling Earth*, Charles Scribner's Sons, 1983, pp. 19–21, 23; Thomas C. Blackburn, ed., *December's Child*, University of California Press, 1975, p. 91; George M. Foster, "A Summary of Yuki Culture," *University of California Anthropological Records*, vol. 5, no. 3 (1944), pp. 207, 233; Edward Curtis, *The North American Indian*, Johnson Reprint Co., 1970, p. 83; Jonathan Barnes, ed., *The Complete Works of Aristotle*, Princeton University Press, 1984, pp. 555–556, 584, 591–597; Jonathan Barnes, ed., *The Cambridge Companion to Aristotle*, Cambridge University Press, 1995, p. 156; and Frank Dawson Adams, *The Birth and Development of the Geological Sciences*, Williams & Wilkins Co., 1938, pp. 399–411, 413.

Native American accounts of earthquakes were drawn from A. L. Kroeber, *Yurok Myths*, University of California Press, 1976, pp. 174–175, 417–418, 460, 464; A. L. Kroeber, "Earthquakes," *Journal of American Folk-lore*, vol. 19 (1906), pp. 322–325;

"Evidence of Giant Quake in 1700s," *San Francisco Chronicle*, November 28, 1995; Alan R. Nelson et al., "Radiocarbon Evidence for Extensive Plate-Boundary Rupture about 300 Years Ago at the Cascadia Subduction Zone," *Nature*, November 23, 1995; Jean Perry, Humboldt State University, August 21, 1996; A. L. Kroeber, "Notes on California Folk-lore," *Journal of American Folk-lore*, vol. 21 (1908), p. 40; James F. Downs, "Washo Religion," vol. 16, no. 9, *University of California Anthropological Records*, p. 367.

3. THE OLD AND NEW WORLDS (1580–1812), PAGES 17–32

London

Oxford English Dictionary, 2nd ed., 1989; Abraham Fleming, "A Bright Burning Beacon, forewarning all Wife Virgins to trim their lamps against the coming of the Bridegroom," 1580; Leet, 1948, p. 19; William Shakespeare (John E. Hankins, ed.), *Romeo and Juliet*, Penguin Books, 1970, pp. 16, 42; "Perpetual and Naturall Prognostications of the Change of Weather," translated from the Italian, printed by John Wolfe, London, 1591; Charles Davison, *A History of British Earthquakes*, Cambridge University Press, 1924, pp. 1, 334.

Lisbon

T. D. Kendrick, *The Lisbon Earthquake*, Methuen & Co., 1956, pp. 87–92, 102–109; Charles F. Richter, *Elementary Seismology*, W. H. Freeman, 1958, pp. 104–105; Officer, 1993, pp. 52–53, 55–58; Harry Fielding Reid, "The Lisbon Earthquake of November 1, 1755," BSSA, June 1914; Charles Davison, *Great Earthquakes*, Thomas Murby and Co., 1936, pp. 1–28; Golden, 1983, pp. 26–30; Ben Ray Redman, *The Portable Voltaire*, Penguin Books, 1977, pp. 241, 565, 567; David Walford, ed., *Immanuel Kant: Theoretical Philosophy, 1755–1770*, Cambridge University Press, 1992, pp. xxxvi–xxxix, lii, lvi, 147; Tributsch, 1982, pp. 15, 137–138, 151; Augustin Udías and William Stauder, "The Jesuit Contribution to Seismology," SRL, May-June 1996.

New England

William T. Brigham, "Volcanic Manifestations in New England," *Memoirs of the Boston Society of Natural History*, vol. 2 (April 7, 1869); Stover, 1993, pp. 250–251, 314; Frederick E. Brasch, "An Earthquake in New England during the Colonial Period (1755)," BSSA, vol. 6 (1916), pp. 26–42; John Mitchell, "Conjectures Concerning the Cause and Observations upon the Phenomena of Earthquakes," read to the Royal Society, 1760; Adams, 1938, pp. 415–419; George P. Merrill, *The First One Hundred Years of American Geology*, Hafner Publishing, 1964, pp. 8–10; Reid, 1914, pp. 62–63; "Earthquakes the Works of God, and Tokens of His Just Displeasure," Thomas Prince, a pastor at South Church, Boston, December 5, 1755.

New Madrid

The many sources of information on the New Madrid quakes include Arch C. Johnston and Lisa R. Kanter, "Earthquakes in Stable Continental Crust," *Scientific American*, March 1990; A. C. Johnston et al., eds., "The Earthquakes of Stable Continental Regions," Electric Power Research Institute, vol. 1, Palo Alto, Calif., 1994, pp. 3/53–3/73; David Lindley, "Not Just in California . . . ," *Nature*, November 23, 1989; Arch C. Johnston, "A Major Earthquake Zone on the Mississippi," *Scientific American*, April 1982; James Lal Penick Jr., *The New Madrid Earthquakes*, University of Missouri Press, 1981; USGS, F. S. McKeown and L. C. Pakiser, *Investigations of the New Madrid, Missouri, Earthquake Region*, Washington, D.C., 1982; Otto W. Nuttli, "The Mississippi Valley Earthquakes of 1811 and 1812: Intensities, Ground Motion and Magnitudes," BSSA, February 1973, p. 227; Kentucky Geological Survey, Ronald L. Street and Otto W. Nuttli, *The Great Central Mississippi Valley Earthquakes of 1811–1812*, Special Publication 14, series 11, 1990; Stover, 1993, pp. 6, 66, 115, 262; "Earthquakes," Kaye M. Shedlock and Louis C. Pakiser, USGS, Washington, D.C., 1994, pp. 2, 14–15; Richard Monastersky, "The Great American Quakes," *Science News*, June 9, 1996; Arch C. Johnston and Kaye M. Shedlock, "Overview of Research in the New Madrid Seismic Zone," SRL, July-September 1992; Maritita P. Tuttle and Eugene S. Schweig, "Archeological and Pedological Evidence for Large Prehistoric Earthquakes in the New Madrid Seismic Zone, Central United States," *Geology*, March 1995; Roy Van Arsdale, "Hazard in the Heartland," *Geotimes*, May 1997; "The Mississippi Valley—'Whole Lotta Shakin' Goin' On,'" USGS Fact Sheet 168-95, 1995; "Uncovering Hidden Hazards in the Mississippi Valley," USGS Fact Sheet 200-96; William Atkinson, *The Next New Madrid Earthquake*, Southern Illinois University Press, 1989; *New York Post*, February 8, 1812; *York Gazette*, January 24, 1812; James Morton Smith, *The Republic of Letters*, W. W. Norton, 1995, p. 1687; "Earthquake," *Louisiana Gazette*, December 21, 1811; "Signs of the Times," *Connecticut Mirror*, April 10, 1812; Robert Penn Warren, *Brother to Dragons*, Louisiana State University Press, 1996; Boynton Merrill Jr., *Jefferson's Nephews*, University Press of Kentucky, 1987.

For water-related aspects I consulted Charles Joseph Latrobe, *The Rambler in North America*, R. B. Seeley and W. Burnside, 1836; Reuben Gold Thwaites, *Early Western Travels*, vol. 5, Arthur H. Clark, 1904, unpaged preface, pp. 201–208, 211;

"Important Arrival," *Louisiana Gazette,* February 22, 1812; "Earthquake," *Pittsburgh Gazette,* February 14, 1812; "Earthquake," *Lexington Reporter,* March 18, 1812; "The Earthquake," *New York Post,* February 11, 1812.

For land-related aspects I drew from Alice Ford, *John James Audubon: A Biography,* Abbeville Press, 1988, pp. 86–87; Maria R. Audubon, *Audubon and His Journals,* Charles Scribner's Sons, 1897, pp. 234–237; Tributsch, 1984, pp. 149, 155; Timothy Flint, *Recollections of the Last Ten Years,* Alfred A. Knopf, 1932, p. 218–219; USGS, Myron L. Fuller, *The New Madrid Earthquake,* Bulletin 494, Washington, D.C., 1912.

4. FORT TEJON (1857), PAGES 33–73

Pallett Creek

John J. Nance, *On Shaky Ground,* William Morrow, 1988, pp. 189–202; E. Joan Baldwin et al., eds., "San Andreas Fault: Cajon Pass to Wallace Creek," South Coast Geological Society, *Guidebook,* vol. 2, no. 17 (1989), p. 9; Ross S. Stein, "Characteristic or Haphazard?" *Nature,* November 30, 1995, p. 443; James P. McCalpin, *Paleoseismology,* Academic Press, 1996; Rick Gore, "Living with California's Faults," *National Geographic,* April 1995, pp. 11, 19, 22; Kerry E. Sieh, "Prehistoric Large Earthquakes Produced by Slip on the San Andreas Fault at Pallett Creek, California," *JGR,* August 10, 1978; Kerry Sieh et al., "A More Precise Chronology of Earthquakes Produced by the San Andreas Fault in Southern California," *JGR,* January 10, 1989; interview with Kerry Sieh, Pasadena, February 13, 1996; Hiroo Kanamori, "Preparing for the Unexpected," *SLR,* January-February 1995, pp. 7–8; Hiroo Kanamori, Gutenberg Lecture: "Waves in the Earth," American Geophysical Union, fall meeting, 1996; interview with Hiroo Kanamori, Caltech, February 4, 1997.

"A Horrifying Earthquake"

Miguel Costansó, *The Narrative of the Portolá Expedition of 1769–1770,* ed. Adolph Van Hemert-Engert and Frederick J. Teggart, Publication of the Academy of Pacific Coast History, vol. 1, no. 4 (March 1910), University of California Press; Miguel Costansó, *Diary of Miguel Costansó,* ed. Frederick J. Teggart, Publication of the Academy of Pacific Coast History, vol. 2, no. 4 (August 1911), University of California Press; Gaspar de Portolá, *Diary of Gaspar de Portolá during the California Expedition of 1769–1770,* ed. Donald Eugene Smith and Frederick J. Teggart, Publication of the Academy of Pacific Coast History, vol. 1, no. 3 (October 1909), University of California Press; Herbert E. Bolton, *Fray Juan Crespi,* University of California Press, 1927; Francisco Palou, *Historical Memoirs of New California,* vol. 2, ed. Herbert E. Bolton, University of California Press, 1926; F. Boneu Companys, *Gaspar de*

Portolá, trans. and ed. Alan K. Brown, Lerida, Spain, 1983, p. 171; Harry O. Wood, "California Earthquakes," *BSSA,* June-September 1916, p. 107.

From Saint Andrew's Abbey to Wrightwood

Interview with Father Vincent and Father Werner, Saint Andrew's Abbey, Valyermo, April 12, 1996; Ruth Reeder, *The Ranch,* self-published, 1994; "Pax," Saint Andrew's Abbey, 1981; *Saint Andrew's Abbey, a Benedictine Monastery,* Valyermo, undated. Information on the low-level flights came from Father Werner; Daniel Sidwell, Table Mountain Facility, Jet Propulsion Laboratory, April 11, 1996; Jim Bowman, Three Points, April 14, 1966; Frankie Sanchez, Pine Mountain, April 15, 1996; Dennis McGrath, public information officer, Lemoore Naval Air Station, May 7, 1996; Capt. Glenn Rogers, Marine Squadron VNFA-121, Miramar Naval Air Station, May 8, 1996; John Jeronomyo, Lemoore Naval Air Station, April 14, 1996; Aldous Huxley, "The Desert," in Leonard Michaels et al., eds., *West of the West,* North Point Press, 1989, p. 322.

For the journey from Valyermo to Big Pines I consulted Paul Greenstein et al., *Bread and Hyacinths: The Rise and Fall of Utopian Los Angeles,* California Classic Books, 1992, pp. 87–97, 116–117; Mike Davis, *City of Quartz,* Vintage, 1992, pp. 9–11; *Earthquake Fault Tour,* Forest Service, Angeles National Forest, Valyermo Ranger District, unpaged and undated; Pearl Comfort Fisher, *The Mountaineers,* self-published, 1972, p. 58; Baldwin, 1989, p. 6.

On the earth wave observed from Table Mountain see "Table Mountain Observatory," Jet Propulsion Laboratory, October 1986; Tufty, 1969, pp. 25–26; Harry O. Wood, "Field Investigation of Earthquakes," *Bulletin of the National Research Council,* October 1933, pp. 61–63; "Seen Ground Waves," sci.geo.earthquakes thread, October 1996; Sidwell, April 11, 1996.

For the Wrightwood area I consulted Robert P. Sharp and Laurence H. Nobles, "Mudflow of 1941 at Wrightwood, Southern California," *GSAB,* May 1963; California Department of Water Resources, Southern Division Report, *Wrightwood Debris and Mud Flow Investigation,* Los Angeles, September 1976, pp. iii, 9; "Earthquake Fault Tour," Forest Service; "Earthquake Geology San Andreas Fault System Palm Springs to Palmdale," Southern California Section Association of Engineering Geologists, 1992; interviews with Frances Yarnell, Wrightwood, April 11, 1996, and Jackie Devlin, Devlin Realty, Wrightwood, April 11, 1996; *The WPA Guide to California,* Pantheon Books, 1984, pp. 614–615; Association of Engineering Geologists, 1992, p. 9; Gordon C. Jacoby Jr. et al., "Irregular Recurrence of Large Earthquakes along the San Andreas Fault: Evidence from Trees," *Science,* July 8, 1988; Ray J. Weldon, "San Andreas Fault, Cajon Pass, Southern California," in *Centennial Field Guide,* vol. 1, Cordilleran Section of the Geological Society of America, 1987, pp. 193–198; Ray J. Weldon and Kerry E. Sieh, "Holocene Rate of Slip and Tentative Recurrence Interval for Large Earthquakes on the San Andreas Fault, Cajon Pass, Southern California," *GSAB*, June 1985; Gary Earney, San Bernardino National Forest special uses team leader, Lytle Creek Ranger Station, May 21, 1996.

The 1812 Overture Continues

Hubert Howe Bancroft, *History of California*, vol. 2, History Co., 1886, p. 200; J. B. Trask, "On the Direction and Velocity of the Earthquake in California on January 9, 1857," *American Journal of Science*, series 2, vol. 25 (January 1858), p. 146; John B. Trask, A *Register of Earthquakes in California: From 1800 to 1863*, Towne and Bacon, 1864, pp. 5, 7; Henry L. Oak, A *Visit to the Missions of Southern California in February and March, 1874*, Southwest Museum, 1981, p. 46; "Mission Life at San Juan Capistrano," KM Communications, 1991, p. 38; *San Francisco Bulletin*, March 5, 1864; Duncan Carr Agnew, "Environmental Perils San Diego Region," *San Diego Association of Geologists*, October 1991, p. 4.1; Harry Kelsey, "San Juan Capistrano Mission Chapels and Cemetery," Padre Press, 1993, pp. 3–4, 53–54; Zephyrin Engelhardt, *San Juan Capistrano Mission*, Mission Museum Press, 1993; Francis J. Weber, *The Jewel of the Missions*, Mission Museum Press, 1992; Angustias de la Guerra Ord, *Occurrences in Hispanic California*, Academy of American Franciscan History, 1954, p. 26; California Division of Mines and Geology, T. R. Toppozada et al., *Preparation of Isoseismal Maps and Summaries of Reported Effects for Pre-1900 California Earthquakes*, Sacramento, 1981, pp. 134–140; Sidney D. Townley and Maxwell W. Allen, "Earthquakes in California, 1769–1928," *BSSA*, vol. 29 (1939), pp. 22–23; Wood, 1916, pp. 108–109.

From Palmdale to Tejon Pass

Allan G. Barrows, "Roadcut Exposure of the San Andreas Fault Zone along the Antelope Valley Freeway near Palmdale, California," Hill, 1987, pp. 211–212; Drew P. Smith, "Roadcut Geology in the San Andreas Fault Zone," *California Geology*, May 1976, pp. 99–104; California Division of Mines, William E. Ver Planck, *Gypsum in California*, San Francisco, September 1952, pp. 57, 67, 70; Baldwin, 1989, pp. 19, 20–24; "AV Incorrectly Perceived as Blighted, Isolated," *Antelope Valley Press*, April 3, 1994; "Prospects Look Bleak for California City," *Wall Street Journal*, February 11, 1994; "Quake Changed Valley Lives," *Antelope Valley Press*, December 28, 1994; "Palmdale 2nd Fastest-Growing City in Country," *Antelope Valley Press*, October 3, 1995; California Division of Mines and Geology, Allen G. Barrows et al., *Geology and Fault Activity of the Palmdale Segment of the San Andreas Fault Zone,*

Los Angeles County, Open-File Report 76-6LA, 1976; "General Plan," City of Palmdale, 1993, pp. S-26, S-36; Barrows, 1976, p. 6; Davis, 1992, p. 12.

For information on the Saint Francis Dam disaster and the stroll through the aqueduct following the 1994 quake see William L. Kahrl, *Water and Power,* University of California Press, 1983, pp. 305, 312–313; Margaret Leslie Davis, *Rivers in the Desert,* HarperPerennial, 1994, pp. 148–151; 171, 193, 223, 251; Royce B. Nunnis Jr., *The St. Francis Dam Disaster Revisited,* Historical Society of Southern California and Ventura County Museum of History and Art, 1995; David Starr Jordan, *The California Earthquake of 1906,* A. M. Robertson, 1907, p. 201; Charles F. Outland, *Man-made Disaster,* Arthur H. Clark, 1977; interview with Dan Kott, Jim Thomas, Carlos Gomez, San Francisquito Canyon Powerhouse No. 1, April 16, 1996.

Three Points and the Neenach Formation were described by Bowman, April 14, 1996; Vincent Matthews, "Correlation of Pinnacles and Neenach Volcanic Formations and Their Bearing on San Andreas Fault Problem," *The American Association of Petroleum Geologists Bulletin,* December 1976, pp. 2128–2141; Tor H. Nilsen, "Offsets along the San Andreas Fault of Eocene Strata from the San Juan Bautista Area and Western San Emigdio Mountains, California," *GSAB,* May 1984, p. 599; Powell, 1993, pp. xii, 211–212; Baldwin, 1989, pp. 11–13, 28; Halka Chronic, *Pages of Stone,* The Mountaineers, 1986, pp. 126–128; Peter W. Weigand and June M. Thomas, "Middle Cenozoic Volcanic Fields Adjacent to the San Andreas Fault, Central California: Correlation and Petrogenesis," *South Coast Geological Society Guidebook,* vol. 1, no. 17 (1990), pp. 207–222; T. W. Dibblee Jr., "Geology of the Devil's Punchbowl, Los Angeles County, California," *Geological Society of America Centennial Field Guide,* 1987, pp. 207–210; interview with David Numer, Devil's Punchbowl, February 14, 1996. (See also notes for Pinnacles National Monument in the Parkfield chapter.)

For Tejon Pass and ranch see "Driving Down Yesterday's Highways, *Pine Mountain Pioneer,* April 8, 1996. Interview with Jim Drago, Caltrans, Sacramento, May 17, 1996; Karl V. Steinbrugge and Donald F. Moran, "An Engineering Study of the Southern California Earthquake of July 12, 1952, and Its Aftershocks," *BSSA,* April 1954, pp. 210–211, 242, 281, 331, 337; Stover, 1993, pp. 102, 144; Richter, 1958, pp. 519–523; Gary J. Onyshko, Tejon Ranch Company, April 15, 1996; Wallace, 1990, pp. 158–159.

Like a Badger

Helen S. Giffen and Arthur Woodward, *The Story of El Tejon,* Dawson's Book Shop, 1942, pp. 75, 83–86; Ann Zwinger, *John Xántus: The Fort Tejon Letters 1857–1859,* University of Arizona Press, 1986, pp. vii–ix, 12–14; Duncan Carr Agnew and Kerry E. Sieh, "A Documentary Study of the Felt Effects of the Great California Earthquake of 1857," BSSA, December 1978 (see also the microfiche appendix); Kerry E. Sieh, "Central California Foreshocks of the Great 1857 Earthquake," BSSA, December 1978; Harry O. Wood, "The 1857 Earthquake in California," BSSA, January 1955; Stover, 1993, pp. 101–102; Wallace, 1990, pp. 158–159; Lynn R. Sykes and Leonardo Seeber, "Great Earthquakes and Great Asperities, San Andreas Fault, Southern California," Geology, December 1985, pp. 835–836; Andrew C. Lawson, The California Earthquake of April 18, 1906, Carnegie Institution of Washington, 1908, pp. 41, 449–451; "The Language of the Salinan Indians," University of California Publications in American Archeology and Ethnology, vol. 14 (1918), p. 120; Philip L. Fradkin, The Seven States of California, University of California Press, 1997, pp. 300–302, 366–367; Muir Dawson, "Southern California Newspapers 1851–1876," Historical Society of Southern California Quarterly, March 1950, pp. 20, 23; T. R. Toppozada et al., "Earthquake History of California," California Geology, February 1986, p. 27; Toppozada, 1981, p. 21; Robert E. Wallace, "Structure of a Portion of the San Andreas Rift in Southern California," GSAB, April 1949, p. 782; Hill, 1981, pp. 112–114; Trask, 1864, p. 4.

Wallace Creek

Blackburn, 1975, p. 91; Myron Angel, Painted Rock, Padre Production, 1979, foreword; Angus MacLean, Cuentos, Pioneer Publishing, undated, p. 46; Wood, 1955, pp. 61–64; J. R. Arrowsmith, "Coupled Tectonic Deformation and Geomorphic Degradation along the San Andreas Fault System," Ph.D. dissertation, Stanford University, 1995; Lisa B. Grant and Andrea Donnellan, "1855 and 1991 Surveys of the San Andreas Fault: Implications for Fault Mechanics," BSSA, April 1994, p. 244; Two Self-Guided Geologic Auto Tours, Carrizo Plain Natural Area, undated pamphlet; Robert E. Wallace, "Notes on Stream Channels Offset by the San Andreas Fault, Southern Coast Ranges, California," in Proceedings of Conference on Geologic Problems of San Andreas Fault System, Stanford University Publication, Geological Sciences, vol. 9 (May 1968); Robert E. Wallace, "The San Andreas Fault in the Carrizo Plain–Temblor Range Region, California," in San Andreas Fault in Southern California, California Division of Mines and Geology, Special Report 118, 1975; Kerry E. Sieh and Richard H. Jahns, "Holocene Activity of the San Andreas Fault at Wallace Creek, California," GSAB, August 1984, pp. 883–896; Wallace, 1990, pp. 94–95; Kerry Sieh and Robert E. Wallace, "The San Andreas Fault at Wallace Creek, San Luis Obispo County, California," in Hill, 1987, pp. 233–238; Lisa B. Grant and Kerry Sieh, "Stratigraphic Evidence for Seven Meters of Dextral Slip on the San Andreas Fault during the 1857 Earthquake in the Carrizo Plain," BSSA, June 1993; Lisa B.

Grant and Kerry Sieh, "Paleoseismic Evidence of Clustered Earthquakes on the San Andreas Fault in the Carrizo Plain, California," *JGR*, April 10, 1994.

5. HAYWARD (1868), PAGES 74–97

Mark Twain and the First Great San Francisco Quake

Mark Twain, *Roughing It*, Rinehart & Co., 1953, pp. x–xi, 315–318; Charles Neider, ed., *The Autobiography of Mark Twain*, HarperPerennial, 1990, pp. 119–122; Bernard Taper, *Mark Twain's San Francisco*, McGraw-Hill, 1963, pp. xxi, 124–128; Edgar Marquess Branch, ed., *The Mark Twain Letters*, vol. 1, University of California Press, 1988, pp. 323–324; Richard B. Lyttle, *Mark Twain*, Atheneum, 1994, pp. 69–73; Stover, 1993, p. 104; John Ripley Freeman, *Earthquake Damage and Earthquake Insurance*, McGraw-Hill, 1932, pp. 167–173; Lawson, 1908, pp. 448–449; Walter L. Huber, "San Francisco Earthquakes of 1865 and 1868," *BSSA*, December 1930, pp. 261–265.

Birth of the Policy of Assumed Indifference

Lawson, 1908, pp. 434–448; Charles Wollenberg, "Life on the Seismic Frontier," *California History*, winter 1992-1993; Freeman, 1932, pp. 173–186; Huber, 1930, pp. 265–272; Stover, 1993, p. 104; California Division of Mines and Geology, Earl W. Hart, ed., *Conference on Earthquake Hazards in the Eastern San Francisco Bay Area*, Special Publication 62, Sacramento, Calif., pp. 321–328; "Prof. Davidson Talks," *San Francisco Bulletin*, April 1, 1898; William H. Prescott, "Circumstances Surrounding the Preparation and Suppression of a Report on the 1868 California Earthquake," *BSSA*, December 1982; Michael L. Smith, *Pacific Visions*, Yale University Press, 1987; Michele L. Aldrich et al., "The 'Report' of the 1868 Haywards Earthquake," *BSSA*, February 1986; Thomas Rowlandson, "A Treatise on Earthquake Dangers, Causes and Palliatives," San Francisco, 1869; *Mining and Scientific Press*, April 28, 1906; J. D. Whitney, "Earthquakes," *North American Review*, April 1869, p. 608; G. K. Gilbert, "Earthquake Forecasts," *Science*, January 22, 1909, pp. 135–136.

The Hayward Fault and the Policy of Assumed Indifference Today

"Map of Recently Active Traces of the Hayward Fault, Alameda and Contra Costa Counties, California," James J. Lienkaemper, USGS Map MF-2196, 1992; USGS, Working Group on California Earthquake Probabilities, *Probabilities of Large Earthquake in the San Francisco Bay Region, California*, USGS Circular 1053, 1990; *On Shaky Ground*, Association of Bay Area Governments, April 1995, pp. 4–5;

"No Fear on the Fault Line," *San Francisco Chronicle*, October 14, 1997; "The Next Big Earthquake," USGS, 1994, pp. 14–15; Sue Ellen Hirschfeld, professor of Geological Sciences, California State University, Hayward, lecture at USGS, Menlo Park, June 27, 1996; "Big Demand for Maps That Forecast Quake Damage," *San Francisco Chronicle*, May 6, 1995; Wollenberg, 1992-1993, p. 507; "Living on the Fault: A Field Guide to the Visible Evidence of the Hayward Fault," Governor's Office of Emergency Services, undated; Douglas S. Dreger et al., eds., "Bulletin of the Seismographic Stations of the University of California at Berkeley," 1992, p. 53; J. J. Litehiser, *Observatory Seismology*, University of California Press, 1989, pp. 25, 33–34, 48, 24–50; George D. Louderback, "History of the University of California Seismographic Stations and Related Activities," BSSA, July 1942, pp. 205–229; Perry Byerly, *Seismology*, Prentice-Hall, 1942, p. 53; John McPhee, *Assembling California*, Farrar, Straus & Giroux, 1993, p. 130; California Seismic Safety Commission, *Planning for the Next One*, Sacramento, Calif., 1991, pp. 189–190, 193, 618–620; California Seismic Safety Commission, *A Compendium of Background Reports on the Northridge Earthquake*, Sacramento, Calif., 1994, pp. 221–223; "Seismic Improvement Study," vol. 1, Earth Sciences Building, University of California, Berkeley, October 19, 1990; *Seismic Safety Correction, Earth Sciences Building*, Major Capital Improvement Program, University of California, Berkeley, June 1992; "UC May Sue over Seismic Work at Quake Lab," *San Francisco Chronicle*, January 9, 1995; "UC's Difficulties in Quake Repairs," *San Francisco Chronicle*, January 21, 1995; Stephen A. Mahin, professor of structural engineering, to Leroy Bean, assistant vice chancellor, Planning, Design and Construction, November 23, 1993; numerous documents dating from November 23, 1993, to June 6, 1994, including communications between Jeffrey S. Gee, senior architect, University of California, and Efren Gutierrez, architect, and Craig Huntington, structural engineer; David R. Stoddart, chair of the Department of Geography, to William Simmons, dean, Social Studies Division, College of Letters and Science, February 15, 1994; Barbara Romanowicz, director of the Seismological Stations, to Thomas Koster, director, Space Management and Capital Programs, January 1, 1994; Chancellor Chang-Lin Tien to Professor William B. N. Berry, Department of Geology and Geophysics, February 4, 1994; G. H. Kimball, Office of Architects and Engineers, University of California, to Dorothy H. Radbruck, USGS, Menlo Park, June 8, 1967; USGS, Dorothy H. Radbruck, *Tectonic Creep in the Hayward Fault Zone*, Geological Survey Circular 525, Washington, D.C., 1996, pp. 3–6; Herbert B. Foster to Luther A. Nichols, comptroller of the University of California, October 5, 1938; George D. Louderback to Luther A. Nichols, preliminary report, June 14, 1939; George D. Louderback to Luther A. Nichols, final report, December 13, 1939; *Living on the Fault*, Governor's Office of Emergency Services, undated; Charles C. Thiel Jr. et al., "State Response to the Loma Prieta Earthquake: Competing against Time," BSSA, October 1991, pp. 2133, 2136–2138, 2140; Jeffrey

Gee, University of California, senior architect, June 17, 1996; California Seismic Safety Commission, 1991, pp. 619–620; University of California at Berkeley Seismograph Station Web site, June 1996.

A *Slowly Evolving Science*

Whitney, 1869, pp. 578–610; G. K. Gilbert, "A Theory of the Earthquakes of the Great Basin, with a Practical Application," *Salt Lake Tribune*, September 20, 1883; Gilbert, 1909, p. 134; Frank Dawson Adams, *The Birth and Development of the Geological Sciences*, Williams & Wilkins, 1938, pp. 420–423; Richter, 1958, pp. 3–4; Charles Davison, *The Founders of Seismology*, Arno Press, 1987; Charles Davison, *A Study of Past Earthquakes*, Charles Scribner's Sons, 1905, pp. 7–8, 21, 321–322; Charles Darwin, *Journal of Researches into the Natural History and Geology of the Countries Visited during the Voyage of H.M.S. "Beagle" Round the World, Under the Command of Capt. Fitz Roy, R.N.*, D. Appleton and Co., 1896, p. 302; Capt. Robert Fitzroy, *Narrative of the Surveying Voyages of His Majesty's Ships "Adventure" and "Beagle,"* Henry Colburn, 1839, pp. 402–404, 406, 410, 414; Davison, 1936, pp. 89–95; Yeats, 1997, pp. 369–370; Bolt, 1993, p. 14; Tributsch, 1982, pp. 51, 165; Charles Lyell, *Principles of Geology*, vol. 1, University of Chicago Press, 1990, pp. xxix, 440–442; Sir Charles Lyell, *A Second Visit to the United States of North America*, John Murray, 1849, pp. 226, 267; Vogel, 1995, pp. 27, 137; Robinson, 1994, p. 57; W. A. Berggren and John A. Van Couvering, eds., *Catastrophes and Earth History*, Princeton University Press, 1984, pp. 4–5, 13, 27, 35; Ager, 1995, pp. xi, xix, 122, 197; Claude C. Albritton Jr., *Catastrophic Episodes in Earth History*, Chapman and Hall, 1989, pp. 44–45, 175–176; Ernest Mayr, *The Growth of Biological Thought*, Harvard University Press, 1982, pp. 362–375; LeGrand, 1994, pp. 2, 12, 17–23; Smith, 1987, pp. 2–4, 29, 42–45, 126; Alan E. Leviton et al., "Frontiers of Geological Exploration of Western North America," Pacific Division, American Association for the Advancement of Science, 1982, pp. 37–66; Patrick L. Abbott and William J. Elliott, "Environmental Perils San Diego Region," *San Diego Association of Geologists*, 1991, p. 80; Joseph LeConte, "Earthquakes," *University Echo*, April–May 1872; Stover, 1993, p. 105; Richter, 1958, pp. 499–503; J. D. Whitney, "The Owens Valley Earthquake," *Overland Monthly*, August and September 1872, pp. 130–140, 266–278; Gilbert, 1883; Davison, 1936, pp. 96–103.

New York City and Charleston

Stover, 1993, pp. 314–316; *New York Herald*, August 11, 12, 13, 1884; *New York Times*, August 11, 12, 1884; Robert P. Stockton, *The Great Shock*, Southern Historical Press, 1986, pp. 18, 28–29, 56–57, 67–73; USGS, Captain C. E. Dutton, *The*

Charleston Earthquake of August 31, 1886, in Ninth Annual Report of the United States Geological Survey to the Secretary of the Interior 1887–88, Washington, D.C., 1889.

Birth of the Concept of Faulting

Mungo Ponton, *Earthquakes*, T. Nelson & Sons, 1888, p. 200; Toppozada, 1986, pp. 27–33; Ellen T. Drake, ed., *Geologists and Ideas: A History of North American Geology*, Geological Society of America, 1985, p. 472; Louderback, 1942, pp. 205–211; Litehiser, 1989, pp. 24–25; Clarence Edward Dutton, *Earthquakes in the Light of the New Seismology*, John Murray, 1904, pp. 54, 60, 70–71, 259; "The Earthquake," *San Francisco Bulletin*, April 1, 1898; Stover, 1993, p. 112; Jordan, 1907, p. 324; Hill, 1981, pp. 112–116, 128; Wallace, 1949, pp. 782–783; Andrew C. Lawson, "Post Pliocene Diastrophism of Southern California," *Geological Sciences*, vol. 1, University of California Publications, pp. 115–160; Andrew C. Lawson, "Sketch of the Geology of the San Francisco Peninsula, California," *USGS, 15th Annual Report*, 1895, pp. 399–476; Frederick Leslie Ransome, "The Probable Cause of the San Francisco Earthquake," *National Geographic*, vol. 17 (1906), p. 280; Gilbert, 1883; Davison, 1905, pp. 5, 245–246.

6. SAN FRANCISCO (1906), PAGES 98–139

5:12 A.M. and To the North

Jordan, 1907, pp. 18–19, 24–25, 102, 113–114, 119, 215–216, 233; Hansen, 1989, pp. 29–30, 136; Lawson, 1908, pp. 1, 58–59, 64, 69, 71–72, 74, 78, 81–82, 178–179, 192–195, 198–206, 369–372; Bruce A. Bolt, "The Focus of the 1906 California Earthquake," *BSSA*, February 1968; David M. Boore, "Strong-Motion Recordings of the California Earthquake of April 18, 1906," *BSSA*, June 1977; Wallace, 1990, pp. 6–7, 124; G. K. Gilbert to A. C. Lawson, May 21, 1906; Tina Marie Niemi, "Late Holocene Slip Rate, Prehistoric Earthquakes, and Quaternary Neotectonics of the Northern San Andreas Fault, Marin County, California," Ph.D. thesis in geology, Stanford University, 1992a, pp. 24, 68; Leet, 1948, pp. 42–49; California Division of Mines and Geology, Robert Streitz and Roger Sherburne, eds., *Studies of the San Andreas Fault Zone in Northern California*, Sacramento, Calif., 1980, pp. 77, 86; Dewey Livingston, Point Reyes National Seashore historian, personal communication and various documents, November 21, 1995; William Bronson, *The Earth Shook, the Sky Burned*, Doubleday, 1959, pp. 150, 132–134; USGS, Grove Karl Gilbert et al., *The San Francisco Earthquake and Fire of April 18, 1906*, Washington, D.C., 1907; J. Edgar Ross, *The Earthquake in Sonoma County*, Healdsburg, 1906.

From Shelter Cove to Fort Ross

Wallace, 1990, pp. 85, 107, 154; Stover, 1993, p. 114; USGS/University of Nevada map, Susan K. Goter et al., "Earthquakes in California and Nevada," Reno, 1994; Stephen R. Walker, "Seismicity Report for Northern California, the Nation, and the World for the Week of July 11–17, 1996," USGS, Menlo Park, July 19, 1996; Lori A. Dengler et al., "Sources of North Coast Seismicity," *California Geology,* March-April 1992; Bernice Wuethrich, "Cascadia Countdown," *Earth,* October 1995; "Averting Surprises in the Pacific Northwest," USGS Fact Sheet 111-95, 1995; Alan R. Nelson et al., "Radiocarbon Evidence for Extensive Plate-Boundary Rupture about 300 Years Ago at the Cascadia Subduction Zone," *Nature,* November 23, 1995; "Evidence of Giant Quake in 1700s," *San Francisco Chronicle,* November 28, 1995; interviews with a number of Shelter Cove residents, including Mario Machi and Cathy Buck, Shelter Cove, July 15, 1996; Vern and Karen Bonham, *Shelter Cove Realty's Showcase of Coastal Homes and Properties,* undated; Ray Raphael, *An Everyday History of Somewhere,* Island Press, 1980; Mario Machi, *The Gem of the Lost Coast: A Narrative History of Shelter Cove,* Eureka Printing Co., 1984, pp. 30–31; USGS, Robert D. Brown Jr. and Edward W. Wolfe, "Map Showing Recently Active Breaks along the San Andreas Fault between Point Delgada and Bolinas Bay, California," Miscellaneous Geologic Investigations Map I-692, Washington, D.C., 1972; Carol S. Prentice, "The Northern San Andreas Fault: Russian River to Point Arena," Division of Geological Sciences, California Institute of Technology, 170-25, undated; *Earthquake,* North Coast Institute, 1982; Judy Bothwell Findeisen, North Fork of the Gualala River, July 16, 1996; Ted Konigsmark, *Geologic Trips, Sea Ranch,* GeoPress, 1995, p. 33; E. O. Essig et al., *Fort Ross,* Limestone Press, 1991, pp. 7, 16–17, 21; Lawson, 1908, pp. 63–64; Joseph Henry Jackson, *Bad Company,* Harcourt, Brace, 1939, pp. 120–121, 177, 235–236, 335; F. Kaye Tomlin, *Some Background Information on Fort Ross; Chinese Presence and the Coast Highway,* 1986; "The Ranch Era," F. Kaye Tomlin, 1991; Laura Call Carr, *My Life at Fort Ross: 1877–1907,* Fort Ross Interpretive Association, 1987, p. 25.

The Pond

J. Samuel Walker, "Reactor at the Fault: The Bodega Bay Nuclear Plant Controversy, 1958–1964—A Case Study in the Politics of Technology," *Pacific Historical Review,* August 1990; Thomas Wellock, "The Battle for Bodega Bay," *California History,* summer 1992; Lawson, 1908, p. 191; Richard L. Meehan, *The Atom and the Fault,* MIT Press, 1984, pp. 1–20, 37, 39–49; Northern California Association to Preserve Bodega Head and Harbor, *Earthquakes, the Atom, and Bodega Head,* undated.

From Tomales Bay to Bolinas Lagoon

John Grissim, "Something Is Out There," *Estero*, summer 1992; Ron Redfern, *The Making of a Continent*, Times Books, 1986, p. 108; *On Shaky Ground*, 1995, p. 4; Philip Williams, "An Evaluation of the Feasibility of Wetland Restoration on the Giacomini Ranch, Marin County," Wetlands Research Associates, 1993; Niemi, 1992a, pp. 97, 124; John C. Merriam, "Ground Sloths in the California Quaternary," *GSAB*, vol. 2 (1900), pp. 612–613; "Geotechnical Investigation Proposed Fradkin Residence," Earth Science Consultants, San Rafael, Calif., July 20, 1989, pp. 3–5, 16; Risa Palm and Michael E. Hodgson, *After a California Earthquake*, University of Chicago Press, 1992, pp. 11, 65; Brian Hill, "Time to Take Action—Before Disaster Strikes," *Motorland*, September-October 1994, p. 6; "Quake Insurance 'Averaging' Aids Bay Area," *Los Angeles Times*, February 2, 1997; Basic Earthquake Policy, California Earthquake Authority, January 1997; Andy Michael, "Earthquake Insurance," ca.earthquakes newsgroup, October 2, 1996; interviews with Lloyd S. Cluff, chairman, California Seismic Safety Commission, San Francisco, July 25, 1997; Hiroo Kanamori, Caltech, Pasadena, February 4, 1997; and Lucile M. Jones, USGS, Pasadena, February 11, 1997; Lawson, 1908, p. 192; interview with Swami Asitananda, Vedanta Retreat, Olema, Calif., August 3, 1996; Tina M. Niemi and N. Timothy Hall, "Late Holocene Slip Rate and Recurrence of Great Earthquakes on the San Andreas Fault in Northern California," *Geology*, March 1992b, pp. 195–198; interview with Richard Kirschman, Doris Ober, et al., Dogtown, Calif., September 27, 1996.

To the South

L. A. Bauer and J. E. Burbank, "The San Francisco Earthquake of April 18, 1906, as Recorded by the Coast and Geodetic Survey Magnetic Observatories," *National Geographic*, vol. 17 (1906); Lawson, 1908, pp. 92, 95–104, 255–257, 280–281; Jordan, 1907, pp. 25–26, 163, 165–187; Gilbert, 1907, pp. 19, 22; "Seventy Bodies Taken Out from Agnew's Ruins," *San Jose Herald*, April 21, 1906; "Inquest Held Over Bodies of Those Killed at Asylum," *San Jose Herald*, April 20, 1906; Bronson, 1959, pp. 138–145; "A Look on the Brighter Side," *San Jose Herald*, April 20, 1906.

To the East

I depended on the following sources for photographic interpretation: Jordan, 1907; Lawson, 1908; Bronson, 1959; Hansen, 1989; "The Great San Francisco Earthquake: 1906," *The American Experience*, WGBH, 1989; exhibits at the Museum of the City of San Francisco, San Francisco; David Cohen et al., eds., *Fifteen Seconds:*

The Great California Earthquake of 1989, Island Press, 1989; "Quake of '89," KRON-TV, San Francisco, 1989; and Rufus Steele, *The City That Is*, A. M. Robertson, 1909.

For text I consulted Hansen, 1989, pp. 40, 87, 97, 136–138, 151; Lawson, 1908, pp. 220–246; Jordan, 1907, pp. 5, 57–60, 254, 321–324; Mary Austin, "The Temblor," in Jordan, 1907, pp. 341–360; Jack London, "The Story of an Eye-witness," *Colliers*, May 5, 1906; Earle Labor, ed., *Portable Jack London*, Penguin Books, 1994, p. 486; James Hopper, "Our San Francisco," *Everybody's Magazine*, June 1906; James Hopper, "A Stricken City's Days of Terror," *Harper's Weekly*, May 12, 1906; Gladys Hansen, "Who Perished," San Francisco Archives Publication No. 2, 1980; Gladys Hansen, "The San Francisco Numbers Game," *California Geology*, December 1987; Steele, 1909, pp. 24, 35–37, 46; Karl V. Steinbrugge, *Earthquake Hazard in the San Francisco Bay Area: A Continuing Problem in Public Policy*, Institute of Governmental Studies, University of California, 1968, p. 45; Gilbert, 1909, p. 137; Streitz, 1980, pp. 167, 171; Bolt, 1993, pp. 2–6; Ransome, 1906, pp. 280–296.

What Quake?

Christopher Morris Douty, *The Economics of Localized Disasters: The 1906 San Francisco Catastrophe*, Arno Press, 1977, pp. 203, 210–212; Arnold J. Meltsner, "The Communication of Scientific Information to the Wider Public: The Case of Seismology in California," *Minerva*, autumn 1979, pp. 333, 335–336, 338–339; Gilbert, 1907, pp. 58–59; "Dangerous Delays in Retrofitting," *San Francisco Chronicle*, February 26, 1996; Steven Tobriner, professor of architectural history, University of California, lecture: "The Phoenix Rises: Aftermath of the 1906 Earthquake and Fire," Berkeley, November 19, 1996; Bronson, 1959, p. 170; *Sunset*, April, May, June-July, October 1906; Hansen, 1989, pp. 126, 153; Steele, 1909, pp. 11, 24; Will Irwin, *The City That Was: A Requiem of Old San Francisco*, B. W. Huebsch, 1906, p. 47; Paul Pinckney, "Lesson Learned from the Charleston Quake, How the Southern City Was Rebuilt Finer Than Ever within Four Years," *San Francisco Chronicle*, May 6, 1906; Fred G. Plummer to G. K. Gilbert, October 10, 1906; Freeman, 1932, pp. v–vi, 2; Stanley Scott, ed., "California's Earthquake Safety Policy," *Earthquake Engineering Research Center*, University of California, Berkeley, 1993, p. xvi; W. M. Davis, "The Long Beach Earthquake," *The Geographical Review*, January 1934, p. 9; Bruce A. Bolt, "The Seismographic Stations of the University of California, Berkeley," *Earthquake Information Bulletin*, USGS, Washington, D.C., May-June 1977; Louderback, 1942, pp. 211–212; "The Earthquake," *BSSA*, March 1911; Litehiser, 1989, p. 28; Andrew C. Lawson, "Seismology in the United States," *BSSA*, March 1911, pp. 3–4; Gilbert, 1909, pp. 134–136; Hill, 1981, pp. 116, 118, 128–129; J. C. Branner, "Earthquakes and Structural Engineering," *BSSA*, March 1913, pp. 2–3; Davison, 1987, p. 152; Drake, 1985, p. 473; Richter, 1958, p. 467; numerous letters from Gilbert to Lawson, Lawson to Gilbert, and Branner to Lawson and Lawson to Branner, in the A. C. Lawson Collection, 1906, 1907, 1908, Bancroft Library, University of California, Berkeley (also see Appendix A in Carol S. Prentice and David P. Schwartz, "Re-evaluation of 1906 Surface Faulting, Geomorphic Expression, and Seismic Hazard along the San Andreas Fault in the Southern Santa Cruz Mountains," BSSA, October 1991); Ellis L. Yochelson, ed., "The Scientific Ideas of G. K. Gilbert, *Geological Society of America*, Special Paper 183, 1980, pp. 38, 41; Harry O. Wood, "Earthquake Study in Southern California," Washington, D.C., Carnegie Institution of Washington, 1935, p. 1; Steinbrugge, 1968, p. 45; Charles W. Jennings, "New Geologic Map of California," *California Geology*, April 1978, p. 77; Meltsner, 1979, p. 338; California Division of Mines and Geology, Charles W. Jennings, "An Explanatory Text to Accompany the Fault Activity Map of California and Adjacent Areas," Sacramento, Calif., 1994, p. 2; "Gladys Hansen: 90 Years Later, Quake Victims Get Names," *San Francisco Chronicle*, April 14, 1996; interview with Gladys Hansen, Museum of the City of San Francisco, June 6, 1996; Hansen, 1980; Hansen, 1987, pp. 271–274; Erica Y. Z. Pan, *The Impact of the 1906 Earthquake on San Francisco's Chinatown*, Peter Land, 1995; "When Will the Next Great Quake Strike Northern California?" USGS Fact Sheet 094-96, 1996; Bolt, 1993, pp. 5, 270.

7. PARKFIELD (1966), PAGES 140–186

From Hollister to Parkfield

Lawson, 1908, pp. 38–39, 288–289; James Z. McCann, *Aftershock: The Loma Prieta Earthquake and Its Impact on San Benito County*, Seismic Publications, 1990; California Division of Mines and Geology, Thomas H. Rogers, *Geology of the Hollister and San Felipe Quadrangles, San Benito, Santa Clara, and Monterey Counties*, Sacramento, Calif., 1993, pp. 11, 15–16; Office of the San Benito County School Superintendent, "San Benito County Long Ago and Today," 1980; San Benito County Chamber of Commerce, *San Benito County: California's Unspoiled Paradise*, 1996 pamphlet; Thomas H. Rogers, "A Trip to an Active Fault in the City of Hollister," undated photocopy; Thomas H. Rogers et al., "Active Displacement on the Calaveras Fault Zone at Hollister, California," BSSA, April 1971; interview with Jerry Damm, Hollister, March 18, 1997; Charles L. Sullivan, entries in unpublished encyclopedia of wine, April 1, 1977; Karl V. Steinbrugge et al., "Creep on the San Andreas Fault," BSSA, July 1960; Marjorie Pierce, *East of the Gabilans*, Western Tanager Press, 1976, pp. 139–140; interview with Pat DeRose, Cienega, March 18, 1997.

For the Pinnacles see George Vancouver, *A Voyage of Discovery to the North Pacific Ocean and Round the World*, John Stockdale, 1801, pp. 122–124; David B. Kier, *An Introduction to the History of San Benito County*, San Benito County Historical Society, undated, p. 8; USGS, Donald C. Ross, *Possible Correlations of Base-*

ment Rocks across the San Andreas, San Gregorio-Hosgri, and Rinconda-Reliz-King City Faults, California, USGS Professional Paper 1317, 1984, p. 37; Clyde Wahrhaftig, *Geologic Notes on Pinnacles National Monument,* San Francisco Hiking Club, 1990; Benjamin Page and Clyde Wahrhaftig, "San Andreas Fault and Other Features of the Transform Regime," *Geology of San Francisco and Vicinity,* 28th International Geological Congress, Guidebook to Field Trip T-105, 1989, pp. 22–27; Vincent Matthews and Ralph C. Webb, *Pinnacles Geological Trail,* undated pamphlet; "Pinnacles," official map and guide, 1991. (See also notes for Neenach Formation in Fort Tejon chapter.)

On Parkfield I consulted Sieh, 1978, p. 1744; William H. Bakun, "History of Significant Earthquakes in the Parkfield Area," *Earthquakes & Volcanoes,* vol. 20, no. 2 (1988); Donalee Thomason, "Historical Vignettes of the 1881, 1901, 1922, 1934, and 1966 Parkfield Earthquakes," *Earthquakes & Volcanoes,* vol. 20, no. 2 (1988); interview with Duane Hamann, Parkfield, March 20, 1997; Lawson, 1908, p. 40; untitled newspaper story dated August 28, 1996, about Buck Kester on the wall of the Parkfield Cafe; Perry Byerly and James T. Wilson, "The Central California Earthquakes of May 16, 1933, and June 7, 1934," *BSSA,* vol. 25, no. 3 (1935); USGS, Robert D. Brown et al., *The Parkfield-Cholame, California, Earthquakes of June-August, 1966—Surface Geologic Effects, Water-Resource Aspects, and Preliminary Seismic Data,* USGS Professional Paper 579, Washington, D.C., 1967; Gordon B. Oakeshott et al., "Parkfield Earthquakes of June 27–29, 1966, Monterey and San Luis Obispo Counties, California—Preliminary Report," *BSSA,* August 1966; Coast and Geodetic Survey, Jerry L. Coffman, ed., *The Parkfield, California, Earthquake of June 27, 1966,* Washington, D.C., 1966; Robert E. Wallace, interviewed by Stanley Scott, "Earthquakes, Minerals, and Me: With the USGS, 1942–1995," USGS Open-File Report 96-260, 1996, pp. 101, 113.

The Seismic Sciences in the Twentieth Century

Wallace, 1996, p. 112; Robinson, 1993, pp. 67–75; Gilbert, 1883; Gilbert, 1909, pp. 121–138; National Research Council, *Practical Lessons from the Loma Prieta Earthquake,* National Academy Press, 1994, p. 76; James C. Williams, "Earthquake Engineering," *Icon,* vol. 1 (1995); Stanley Scott, ed., "California's Earthquake Safety Policy: A Twentieth Anniversary Retrospective, 1969–1989," Earthquake Engineering Research Center, University of California, Berkeley, 1993, pp. xiii–xxviii; Wood, 1955, pp. 57–58; Agnew and Sieh, 1978, p. 11; Meehan, 1984, p. 57; Thomas H. Heaton, "Urban Earthquakes," *SRL,* September-October 1995, pp. 37, 39; House of Representatives, Committee of Science, Space, and Technology, *Lessons Learned from the Northridge Earthquake,* March 2, 1994, pp. 323–326, 275–276, 279–280, 282, 290, 293, 299; Stanley Scott, ed., "Henry J. Degenkolb," Earthquake Engineering

Research Institute Oral History Series, 1994, p. 132; Governor's Board of Inquiry on the 1989 Loma Prieta Earthquake, Charles C. Thiel Jr., ed., *Competing against Time*, Sacramento, Calif., 1990, pp. 117–120; CSSC, 1994, p. 114; George W. Housner, professor emeritus of earthquake engineering, Caltech, February 3, 1997; Thomas H. Heaton, "Looking Back from the Year 3000," *SRL*, March-April 1995, pp. 3–4; also interviews with Wilfred Iwan, Thomas H. Heaton, Hiroo Kanamori, and Clarence Allen of Caltech; James J. Mori and Lucile M. Jones of the Pasadena office of the USGS (February 1997); and William Ellsworth and David P. Schwartz of the Menlo Park office of the USGS (May 1997).

The section on the evolution of the theory of plate tectonics was drawn from Meltsner, 1979, p. 338; H. E. LeGrand, *Drifting Continents and Shifting Theories*, Cambridge University Press, 1988; James Romm, "A New Forerunner for Continental Drift," *Nature*, February 3, 1994; A. Hallam, *Great Geological Controversies*, Oxford University Press, 1983, pp. 110–156, 131, 135–156; Moores, 1990 and 1996; Gilbert, 1909, pp. 121–123; J. Tuzo Wilson, "Continental Drift," *Continents Adrift*, W. H. Freeman, 1973, pp. 41–55; Officer, 1992, p. 11; William Glen, *Continental Drift and Plate Tectonics*, Charles E. Merrill, 1975; Tanya Atwater, "Implication of Plate Tectonics for the Cenozoic Tectonic Evolution of Western North America," *GSAB*, December 1970; Harry O. Wood, "Earthquakes in California," *Scientific Monthly*, October 1934, pp. 323–344; Hill, 1981, pp. 119–120, 124, 126; interviews with Ellsworth and Schwartz of the USGS Menlo Park office (May 1997), Mori and Jones of the USGS Pasadena Office (February 1997), and Iwan, Heaton, Kanamori, and Allen of Caltech (February 1997); Mason L. Hill and T. W. Dibblee, Jr., "San Andreas, Garlock, and Big Pine Faults, California," *BSSA*, April 1953, pp. 444–445, 448; Kenneth A. Brown, *Cycles of Rock and Water*, HarperCollins West, 1993, pp. 108–109; Wallace, 1949, p. 783; J. B. Macelwane, "Forecasting Earthquakes," *BSSA*, January 1946, pp. 1–4; Vogel, 1995, pp. 21–28, 159–168; Robinson, 1993, pp. 20–30, 57–58; Ager, 1995, p. 197; Thomas S. Kuhn, *The Structure of Scientific Revolutions*, University of Chicago Press, 1970, p. 24.

The Birth of the Modern Prediction Effort

Wallace, 1996, pp. 60, 63, 75, 77, 87–89, 91; Macelwane, 1946; Panel on Earthquake Prediction of the Committee on Seismology, *Predicting Earthquakes*, National Academy of Sciences, Washington, D.C., 1976, pp. 2, 15–16; Cinna Lomnitz, *Fundamentals of Earthquake Prediction*, John Wiley & Sons, 1994, pp. viii–xi, 32–33; Gilbert, 1909, p. 134; Bryce Walker, *Earthquake*, Time-Life Books, 1982, p. 152; Tsuneji Rikitake, *Earthquake Forecasting and Warning*, Center for Academic Publications, 1982, p. 2; interviews with Mori (February 1997) and Ellsworth (May 1997) of the USGS; Gilbert F. White et al., *Assessment of Research on Natural Hazards*,

MIT Press, 1975, p. 330; Committee on Socioeconomic Effects of Earthquake Predictions, A *Program of Studies on the Socioeconomic Effects of Earthquake Predictions*, National Academy of Sciences, Washington, D.C., 1978; Ralph H. Turner, "Earthquake Predictions: Potential Blessing," *California Geology*, September 1976, pp. 208–209; Officer, 1992, pp. 43–45; Dennis S. Mileti et al., *The Great Earthquake Experiment*, Westview Press, 1993, pp. 20–23, 47–48; Robert A. Page et al., *Goals, Opportunities, and Priorities for the USGS Earthquake Hazards Reduction Program*, USGS Circular 1079, Washington, D.C., 1992; Robert L. Wesson and Robert E. Wallace, "Predicting the Next Great Earthquake in California," *Scientific American*, February 1985, p. 35; Duncan Carr Agnew and William L. Ellsworth, "Earthquake Prediction and Long-Term Hazard Assessment," *Reviews of Geophysics*, U.S. International Report to International Union of Geodesy and Geophysics, 1991, pp. 877–889; David Applegate and Murray Hitzman, "White House, Congress Rethink National Strategy," *SRL*, July-August 1996; Office of the President, National Earthquake Strategy Working Group, *Strategy for National Earthquake Loss Reduction*, Strategic Plan, April 1996, pp. 18–19; Scott, 1994, p. 168; interview with Robert E. Wallace, Menlo Park, May 22, 1997; Stanley Scott, ed., "California's Earthquake Safety Policy: A Twentieth Anniversary Retrospective, 1969–1989," Earthquake Engineering Research Center, University of California, Berkeley, December 1993, pp. 63–64; interview with Cluff, July 25, 1997.

The Chinese Experience

Tang Xiren, A *General History of Earthquake Studies in China*, Beijing, China: Science Press, 1988; Robinson, 1993, p. 48; Li Hui, "China's Campaign to Predict Quakes," *Science*, September 13, 1996; Chen Zhangli, "Progress in Earthquake Studies in China for the Past 30 Years," *Earthquake Research in China*, vol. 10, no. 1 (1996), pp. 7–9; Group Responsible for Mass Monitoring and Protection against Earthquake Hazards, "On the Role of the Mass Monitoring and Mass Protection Programme against Earthquake Hazard in China," in *Earthquake Prediction*, Proceedings of the International Symposium on Earthquake Prediction, Terra Scientific Publishing Co., 1984, pp. 869–876; Ma Zongjin, *Earthquake Prediction*, Seismological Press, 1984, pp. vi–9; Rikitake, 1982, pp. 33–44; Haicheng Earthquake Study Delegation, "Prediction of the Haicheng Earthquake," *EOS, Transactions, American Geophysical Union*, April 1977; Lomnitz, 1994, pp. 22–23, 19, 62; Ma Zongjin et al., *Earthquake Prediction: Nine Major Earthquakes in China (1966–1976)*, Springer-Verlag, 1990, pp. vi–vii, 1–6; Chen Yong, *The Great Tangshan Earthquake of 1976*, Pergamon Press, 1988; Lomnitz, 1994, pp. 14–20, 24–31, 61–66; Mileti, 1993, p. 49; Quian Guan, "1976 Tangshan, China, Earthquake," *California Geology*, December 1987; Paul C. Jennings, ed., *Earthquake Engineering and Hazards Reduction in*

China, National Academy of Science, Washington, D.C., 1980; Tributsch, 1982, pp. 23–28; William L. Corliss, *Handbook of Unusual Natural Phenomena*, Arlington House, 1986; "Anomalous Macroscopic Phenomena and Their Significance in the Prediction of Strong Earthquakes," Zhu Feng-ming and Zhong Yi-zhang, *Earthquake Prediction*, 1984, pp. 193–201; "The Quake That Shook Mexico Awake Is Recalled," *New York Times*, September 19, 1995; Alexander, 1993, p. 57; Stover, 1993, p. 75; Nance, 1988, pp. 226–232; "Tangshan Journal," *New York Times*, July 27, 1996.

Earthquake Phenomena

Jennings, 1980, pp. 150–151; Robert E. Wallace and Ta-Liang Teng, "Prediction of the Sungpan-Pingwu Earthquakes, August 1976," *BSSA*, August 1980; Tributsch, 1984, pp. x, 1–121, 132–142; Lyell, 1990, p. 400; John Milne, *Earthquakes*, P. Blakiston's & Sons, 1939, pp. 15–19; J. E. White, *Underground Sound*, Elsevier, 1983, p. vii; Corliss, 1986, pp. 37–38, 272, 272–275, 277–279, 290–295, 296–299, 358–362; David Starr Jordan, "The Earthquake and Professor Larkin," *Science*, August 10, 1906, pp. 178–180; Lawson, 1908, pp. 377–379, 382–383; H. O. Wood, "The Observation of Earthquakes: A Guide for the General Observer," *BSSA*, June 1911, pp. 78–81; Subsidiary Committee on Seismology, "Physics of the Earth—VI Seismology," *Bulletin of the National Research Council*, October 1933, pp. 59–60; "An Earthquake Phenomenon," *Press Democrat* (Santa Rosa), November 30, 1969; John S. Derr, "Earthquake Lights: A Review of Observations and Present Theories," *BSSA*, December 1973; Edward U. Condon, *Scientific Study of Unidentified Flying Objects*, E. P. Dutton, 1969, pp. 740–741; Charles Davison, "Earthquake Sounds," *BSSA*, July 1938; Leet, 1948, pp. 51–52; David P. Hill et al., "Earthquake Sounds Generated by Body-wave Ground Motion," *BSSA*, August 1976; "Earthquake Sounds 'Captured,' " *California Geology*, April 1976, p. 91; Dutton, 1889, p. 322; Al Martinez, *City of Angles*, St. Martin's Press, 1996, p. 26; Barnes, 1984, p. 593; Alexander, 1993, pp. 554–574; "Employee Counseling Sessions to Deal with Emotional Aftermath of Earthquake," Stanford University News Service, November 8, 1989; Bolt, 1993, p. 230; Ralph H. Turner, "Individual and Group Response to Earthquake Prediction," *Earthquake Prediction*, 1984, p. 601; Jeffrey Goodman, *We Are the Earthquake Generation*, Seaview Books, 1976, pp. 202–210; Leon S. Otis, "Biological Premonitors of Earthquakes: A Validation Study," Stanford Research Institute, final report, Menlo Park, Calif., 1985; Leon S. Otis, "Biological Precursors of Earthquakes: A Validation Study," in *Earthquake Prediction*, 1984, pp. 253–262; UCLA, Institute of Geophysics and Planetary Physics, "Can Animals Predict Earthquakes?" annual technical report, Los Angeles, 1979; Dale F. Lott et al., "Unusual Animal Behavior Prior to a Moderate Earthquake," semiannual technical progress report, University of California at Davis, 1979; Ruth B. Simon, "Animal Behavior and Earthquakes,"

California Geology, September 1976, pp. 210–211; Helmut Tributsch, "Do Aerosol Anomalies Precede Earthquakes?" *Nature,* December 7, 1978, pp. 606–607.

Parkfield and the Politics of Prediction

Ding Guo-yu et al., "Methods of Earthquake Prediction," in *Earthquake Prediction,* 1984, pp. 453–465; Carey McWilliams, *Southern California Country,* Peregrine Smith Books, 1988, pp. 250–251; William B. Rice, *William Money,* Los Angeles, privately printed, 1943; Goodman, 1976; John R. Gribbon and Stephen H. Plagemann, *The Jupiter Effect,* Walker, 1974; Curt Gentry, *The Last Days of the Late, Great State of California,* G. P. Putnam, 1968; Mileti, 1993, pp. 53–54; USGS director V. E. McKelvey to Governor Edmund G. Brown Jr., March 3, 1976; H. William Menard, USGS director, to Alex R. Cunningham, director of the State Department of Emergency Services, July 3, 1980; Robert O. Castle et al., "Aseismic Uplift in Southern California," *Science,* April 1976, pp. 251–253; Wallace, 1996, pp. 101, 110–112; Ralph H. Turner et al., *Waiting for Disaster,* University of California Press, 1986; Charles R. Real and John H. Bennett, "Palmdale Bulge," *California Geology,* August 1976; Jack Bennett, "Palmdale 'Bulge' Update," *California Geology,* August 1977; USGS Director Dallas L. Peck to Cunningham, October 1981; Robert L. Wesson, "Procedures for the Evaluation and Communication of Earthquake Predictions within the United States," in *Earthquake Prediction,* 1984, pp. 970–975; Lomnitz, 1994, pp. 34, 35–36, 139, 271–272; Richard Stuart Olson et al., *The Politics of Earthquake Prediction,* Princeton University Press, 1989; General Accounting Office, "Stronger Direction Needed for the National Earthquake Program," GAO 83-103, Washington, D.C., July 26, 1983; John R. Filson, "The Role of the Federal Government in the Parkfield Earthquake Prediction Experiment," *Earthquakes & Volcanoes,* vol. 20, no. 2 (1988), pp. 56–58; Mileti, 1993, pp. 61–85, 101–106; H. M. Iyer in *Earthquake Prediction,* 1986, pp. 1–11; Wallace, May 22, 1997; James H. Dieterich, chief scientist, Western Regions Earthquake Hazards, Menlo Park, Calif., May 21, 1997; Officer, 1992, pp. 12–13; Marcel C. LaFollette, *Stealing into Print,* University of California Press, 1992, pp. 30, 119; William H. Bakun and Thomas V. McEvilly, "Earthquakes near Parkfield, California: Comparing the 1934 and 1966 Sequences," *Science,* September 28, 1979; William H. Bakun and Thomas V. McEvilly, "Recurrence Models and Parkfield, California, Earthquakes," *JGR,* May 10, 1984, pp. 3051–3058; William H. Bakun and Allen G. Lindh, "The Parkfield, California, Earthquake Prediction Experiment," *Science,* August 16, 1985, pp. 619–624; Wallace, 1996, pp. 114–115, 172; National Earthquake Prediction Evaluation Council, Menlo Park, minutes of the November 1984 meeting; National Earthquake Prediction Evaluation Council, Pasadena, minutes of the March 29–30, 1985, meeting; USGS director Dallas Peck to William M. Medigovich, director of the California

Office of Emergency Services, April 4, 1985; USGS, "Studies Forecasting Moderate Earthquake near Parkfield, Calif., Receive Official Endorsement," press release from the USGS Public Affairs Office, Reston, Va., April 5, 1985; Wesson and Wallace, 1985, pp. 35–43; Allen A. Boraiko, "Earthquake in Mexico," *National Geographic*, May 1986, p. 667; USGS, 1988, pp. 42, 60–78, 83–86; Richard D. McJunkin et al., "The Parkfield Strong-Motion Array," *California Geology*, February 1983, pp. 27–34; J. C. Savage, "Criticism of Some Forecasts of the National Earthquake Prediction Evaluation Council," *BSSA*, June 1991, pp. 862–869; J. C. Savage, "The Parkfield Prediction Fallacy," *BSSA*, February 1993, pp. 1–6; interview with James C. Savage, Menlo Park, Calif., May 20, 1997; Lomnitz, 1994, pp. 46–54; interview with Allan G. Lindh, Menlo Park, Calif., July 28, 1997; USGS, National Earthquake Prediction Evaluation Council Working Group, "Earthquake Research at Parkfield, California, 1993 and Beyond," USGS Circular 1116, Washington, D.C., 1994; "Quake Prospecting That Didn't Pan Out," *San Francisco Chronicle*, December 12, 1992; Richard A. Kerr, "Parkfield Quakes Skip a Beat," *Science*, February 19, 1993; William D. Stuart, "Parkfield Slowing Down," *Nature*, May 31, 1990, p. 383; M. Wyss et al., "Decrease in Deformation Rate as a Possible Precursor to the Next Parkfield Earthquake," *Nature*, May 31, 1990, p. 430; "Big Quake Predicted for Parkfield," *San Francisco Chronicle*, November 11, 1993; "Quake Chances Diminishing in Shaky Parkfield," *San Francisco Chronicle*, November 16, 1993; "Media Swarm Quake Country," *San Francisco Chronicle*, November 16, 1993; "Big Quake Unlikely, Scientists Concede," *San Francisco Chronicle*, November 17, 1993; "Scientists Call Off Earthquake Alert Near Parkfield," *San Francisco Chronicle*, November 18, 1993; "Quake Experts Still Support Forecasting Despite Recent Miss," *San Francisco Chronicle*, November 22, 1993; Andrew J. Michael and John Langbein, "Earthquake Prediction Lessons from Parkfield Experiment," *EOS, Transactions, AGU*, March 30, 1993, pp. 145, 153–155; interview with John Langbein, Menlo Park, Calif., May 22, 1997; David D. Jackson, "The Parkfield Probability Problem," *EOS, Transactions, AGU*, spring meeting, 1997, p. S218; "Hopes for Predicting Earthquakes Once So Bright, Are Growing Dim," *New York Times*, August 8, 1995; Robert J. Geller, "Predicting Earthquakes Is Impossible," *Los Angeles Times*, February 2, 1997; Christopher Scholz, "Whatever Happened to Earthquake Prediction?" *Geotimes*, March 1997; Christopher H. Scholz, "Earthquakes as Chaos," *Nature*, November 15, 1990, pp. 197–198; Ross S. Stein, "Characteristic or Haphazard?" *Nature*, November 30, 1995, pp. 443–444.

8. LOMA PRIETA (1989), PAGES 187–231

5:04 P.M. on an Unusual Tuesday

National Research Council, *Practical Lessons from the Loma Prieta Earthquake,* National Academy Press, 1994; Lee Benuska, ed., "Loma Prieta Earthquake Reconnaissance Report," *Earthquake Spectra,* Earthquake Engineering Research Institute, 1990; California Seismic Safety Commission (CSSC), *Loma Prieta's Call to Action,* Sacramento, Calif., 1991; CSSC, *Planning for the Next One: Transcripts of Hearings on the Loma Prieta Earthquake of October 17, 1989,* Sacramento, Calif., 1991, pp. 7, 241; Thomas Y. Canby, "California Earthquake—Prelude to the Big One?" *National Geographic,* May 1990; George Plafker and John P. Galloway, eds., *Lessons Learned from the Loma Prieta, California, Earthquake of October 17, 1989,* USGS Circular 1045, Washington, D.C., 1989; Robert Bolin, ed., *The Loma Prieta Earthquake: Studies of Short-Term Impacts,* Institute of Behavioral Science, University of Colorado, 1990, p. 44; Laura A. Thorpe, ed., *Three Weeks in October,* Woodford Publishing, 1990; "Quake of '89: A Video Journal," KRON-TV video cassette.

On Very Shaky Ground

Charles R. Scawthorn et al., "Performance of Emergency-Response Services after the Earthquake," in Thomas D. O'Rourke, ed., *The Loma Prieta, California, Earthquake of October 17, 1989—Marina District,* USGS Professional Paper 1551-F, Washington, D.C., 1992; "The Marina," *Neighborhood Focus,* Pacific Union Co., 1992; "Streets to Live By," *San Francisco Examiner,* June 11, 1995; California Division of Mines and Geology, *The Loma Prieta (Santa Cruz Mountains) California, Earthquake of 17 October 1989,* Special Publication 104, Sacramento, Calif., 1990, pp. 116–117; CSSC, "Planning for the Next One," 1991, pp. 51, 243; Bolt, 1993, p. 272; Thomas C. Hanks and Helmut Krawinkler, "The 1989 Loma Prieta Earthquake and Its Effects: Introduction to the Special Issue," *BSSA,* October 1991, pp. 1419–1420; M. G. Bonilla, "Geologic and Historical Factors Affecting Earthquake Damage," in O'Rourke, 1992, pp. F9–23, F25; Wallace, 1990, pp. 62–66; Arthur Quinn, *Broken Shore,* Redwood Press, 1987, pp. 3–4; David Van Horn, "Harbor View," in Roger and Nancy Olmsted, "San Francisco Waterfront Report on Historical Cultural Resources for the North Shore and Channel Outfalls Consolidation Projects Prepared for the San Francisco Wastewater Management Program," Technigraphics, 1977, pp. 669–718; Lawson, 1908, pp. 229, 232, 241; J. L. Chameau et al., "Liquefaction Response of San Francisco Bayshore Fills," *BSSA,* October 1991, p. 2015; Plafker, 1989, p. 25; Benuska, 1990, pp. 84–114; James K. Mitchell et al., "Soil Conditions and Earthquake Hazard Mitigations in the Marina District of San

Francisco," Department of Civil Engineering and Earthquake Engineering Research Center, University of California, Berkeley, 1990, pp. 27, 34–35; Kevin Starr, *Americans and the California Dream*, Oxford University Press, 1973, pp. 295–306; Thomas L. Holzer, "Summary and Conclusions," and M. G. Bonilla, "Natural and Artificial Deposits in the Marina District," in *Effects of the Loma Prieta Earthquake on the Marina District of San Francisco, California*, USGS Open-File Report 90-253, Menlo Park, Calif., 1990; M. G. Bonilla, "The Marina District, San Francisco, California: Geology, History, and Earthquake Effects," *BSSA*, October 1991.

"I Knew My Family Was in Trouble"

CSSC, 1991, pp. 22–23, 25, 39–45; various exhibits at the Museum of the City of San Francisco; various USGS reports cited above.

The Media: Like Moths to a Flame

PG&E and the Earthquake of '89, Pacific Gas & Electric Co., 1990; Everett M. Rogers et al., "Mass Media Coverage of the 1989 Loma Prieta Earthquake: Estimating the Severity of a Disaster," Annenberg School for Communication, University of Southern California, July 1990; Everett M. Rogers et al., in Bolin, 1990, pp. 44–53; Don DeLillo, *White Noise*, Penguin, 1984, p. 66; Todd Gitlin, "Gauging the Aftershocks of Disaster Coverage," *New York Times*, November 12, 1989; *Covering the Quake*, Graduate School of Journalism, University of California at Berkeley, December 9, 1989; Students of the Marina Middle School, *Aftershocks*, Marina Middle School, 1990, pp. 94–95; CSSC, 1991, pp. 45, 238, 632; Richard Rapaport, "Shakedown for the Media," *San Francisco Focus*, December 1989; Betty Medsger, "Earthquake Shakes Four Newspapers," *Washington Journalism Review*, December 1989; CSSC, 1991, pp. 170–175, 375–377; Benuska, 1990, pp. 311–314; David Cohen et al., eds., *Fifteen Seconds*, Tides Foundation/Island Press, 1989, p. 82; Conrad Smith, *Media and Apocalypse*, Greenwood Press, 1992, pp. 115–144; Federico Subervi-Vélez et al., *Communicating with California's Spanish-Speaking Populations*, California Policy Seminar, University of California, 1992; Governor's Office of Emergency Services, Linda Bourque et al., *Experiences during and Responses to the Loma Prieta Earthquake*, July 1994, pp. 47–59.

The Hidden Costs of Earthquakes

Brenda D. Phillips, "Living in the Aftermath: Blaming Processes in the Loma Prieta Earthquake," Department of Sociology, Southern Methodist University, draft report, 1990; CSSC, 1991, pp. 82, 96, 302, 309, 312, 323, 334–335, 337, 339,

341–344, 360, 384, 473–474; National Research Council, 1994, pp. 123–124; Governor's Office of Emergency Services, *Earthquake Recovery: A Survival Manual for Local Government*, September 1993, p. 365; Bolin, 1990, pp. 8, 18–19, 103–105; Brenda D. Phillips, "Post-Disaster Sheltering and Housing of Hispanics, the Elderly and the Homeless," final project report, Department of Sociology, Southern Methodist University, Dallas, Tex., 1991; Benuska, 1990, pp. 191–192, 380–381; CSSC, 1991, pp. 89–96; National Research Council, 1994, pp. 23, 124; Bolin, 1990, pp. 103–105.

From Daly City to San Juan Bautista

Matthew R. Clark, "Archeological Investigations of the Mussel Rock Site," Holman & Associates, archeological consultants, San Francisco, 1986; Lawson, 1908, p. 250; "Living on the Fault," Governor's Office of Emergency Services, undated; Samuel C. Chandler, *Gateway to the Peninsula: A History of the City of Daly City*, 1973, pp. 27–28, 107–109; Alan Hynding, *From Frontier to Suburb: The Story of the San Mateo Peninsula*, Star Publishing Co., 1982, pp. 273–276; Gordon B. Oakeshott, ed., *San Francisco Earthquake of March 1957*, California Division of Mines Special Report 57, 1959; Stover, 1993, p. 150; "$378,000 Quake Damage," *Daly City Record*, March 28, 1957; J. E. Baldwin II et al., "Loma Prieta Earthquake: Engineering Geologic Perspectives," Association of Engineering Geologists, Special Publication No. 1, 1991, pp. 79–82; "Summary of Damages," City of Daly City interoffice memo, October 18, 1989; "Quake Update," senior management analyst to city council, November 6, 1989; "It's Their Fault: Living on the Edge in Daly City," *Daly City Record*, November 4, 1989; Safety Element, Daly City General Plan, 1994; "Lot Going for Little Boxes," *San Francisco Examiner*, February 19, 1995; *Community Guide and Business Directory*, Daly City–Colma Chamber of Commerce, 1995; undated map of fault line in files of Daly City Planning Department; Raymond Sullivan et al., "Living in Earthquake Country," *California Geology*, January 1977; M. G. Bonilla and Julius Schlocker, "Guide to San Francisco Peninsula," California Division of Mines and Geology, Field Trip B, Bulletin 190, pp. 441–450; Interview with Art Beighley, Daly City, November 15, 1996; others interviewed on November 12 and 15; Canby, 1992, p. 110; "The San Andreas Transform Belt," Field Trip Guidebook T303, 28th International Geological Congress, Long Beach, Calif., 1989, pp. 108–109; Erwin G. Gudde, *1000 California Place Names*, University of California Press, 1959, p. 69; Warren D. Hanson, *San Francisco Water and Power: A History of the Municipal Water Department and Hetch Hetchy System*, City and County of San Francisco, 1994, pp. 13–14; Streitz, 1980, pp. 173, 181–187; Robert B. Jansen, *Dams and Public Safety*, Water and Power Resources Service, 1980, pp. 237–240; Hansen, 1994, pp. 17–18, 45; Lawson, 1908, p. 99; CSSC, 1991, pp. 411–416, 435, 439, 572, 574–576; "Investigation of Seismic Stability Lower Crystal Springs Dam," draft final report, W. A. Wahler & Associates, 1977, I-1, II-1–3, V-1, V-6–9; William H. Camp, manager of operations engineering, San Francisco Water Department, November 13, 1996; California Division of Safety of Dams, "National Dam Inspection Program, Lower Crystal Springs Dam," Army Corps of Engineers, Sacramento District, September 1980; Warren Hanson, *San Francisco Water and Power,* City and County of San Francisco, 1985.

"Stanford Self-Insured for Earthquake Damage," Stanford University News Service, October 19, 1989; "President Donald Kennedy and Provost James Rosse Respond to Some Commonly Asked Questions about the Impact of the Earthquake at Stanford," Stanford University News Service, October 24, 1989; "Great Earthquake of '06 Cleared Away Unsafe Buildings," Stanford University News Service, October 24, 1989; Lawson, 1908, p. 256; "Guide to Buildings and Exhibits," Stanford University School of Earth Sciences, 1993; "Stanford's Earthquake Expert Tours and Assesses Earthquake-Damaged Buildings," Stanford University News Service, October 19, 1989; "Stanford, FEMA, Settle 1989 Earthquake Damage Claims," Stanford University News Service, March 16, 1994; "Earthquake Risk Management Commission Final Report," Stanford University, November 26, 1990; "Stanford Quad Ready to Come Out from Under Its Wraps," *San Francisco Chronicle,* August 16, 1996; CSSC, 1991, pp. 459–464; Carol S. Prentice and David P. Schwartz, "Re-evaluation of 1906 Surface Faulting, Geomorphic Expression, and Seismic Hazard along the San Andreas Fault in the Southern Santa Cruz Mountains," *BSSA,* October 1991; R. E. Ruland, "A Summary of Ground Motion Effects at SLAC Resulting from the Oct. 17th 1989 Earthquake," Stanford Linear Accelerator Center, August 1990; P. A. Moore, Stanford Linear Accelerator Center, July 30, 1997; "Two Jolts Mark Anniversary of 1906 Quake," *San Francisco Chronicle,* April 19, 1996; "Whole Lotta Shakin' Goin' On," *Measure* (Hewlett-Packard internal publication), January-February 1990; "Bracing for the Big One," *Measure,* July-August 1992, pp. 11–12; CSSC, 1991, pp. 442–450; Bob Lanning, Palo Alto, November 20, 1996.

CSSC, 1991, pp. 8, 360; Plafker, 1989, pp. 9–16; Prentice, 1991, pp. 1424, 1455–1456; David P. Schwartz and Daniel J. Ponti, eds., *Field Guide to Neotectonics of the San Andreas Fault System, Santa Cruz Mountains, in Light of the 1989 Loma Prieta Earthquake,* USGS Open-File Report 90-274, 1990, pp. 6, 23, 28, 30; USGS staff, 1990, pp. 286–293; Working Group on California Earthquake Probabilities, 1990, pp. 14–15; Sandy Lydon, *Chinese Gold,* Capitola Book Co., 1985, pp. 141–142, 154; Rick Hamman, *California Central Coast Railways,* Pruett Publishing Co., 1980, pp. 102, 138, 140–142, 154; Lawson, 1908, p. 111; "Docents' Manual," San Juan Bautista State Park Volunteer Association; *History of San Benito County, California,* 1881, p. 91; Stover, 1993, p. 99; Glenn Farris, state archaeologist II, "Former Mission Building Foundations on the Taix Lot at San Juan Bautista

SHP," 1991; Hubert Howe Bancroft, *History of California*, vol. 1, p. 559; Martha H. Lowman, *California's Mission San Juan Bautista*, undated pamphlet, pp. 2–6; Mary Null Boulé, *Mission San Juan Bautista*, 1988, pp. 12–13; John M. Martin, *Mission San Juan Bautista*, undated and unpaged pamphlet; Fradkin, 1997, pp. 274–277.

Two Sequels: New York and New Madrid

National Research Council, 1994, p. 66–67; C. Thomas Statton, "Seismic Hazard and Design in the New York Metropolitan Area," Woodward-Clyde Consultants, Metropolitan Section of the American Society of Civil Engineering, 1990; Klaus H. Jacob, "Seismic Hazards and the Effects of Soils on Ground Motions for the Greater New York City Metropolitan Area," Lamont-Doherty Geological Observatory, Metropolitan Section of the American Society of Civil Engineering, 1990; L. Seeber and J. G. Armbruster, "A Study of Earthquake Hazards in New York State and Adjacent Areas," Lamont-Doherty Geological Observatory, for the U.S. Nuclear Regulatory Commission, 1986; Fred Graver and Charlie Rubin, "Waiting for the Big One," *New York*, December 11, 1995; Penick, 1981, pp. 152–154; Williams Spence et al., *Responses to Iben Browning's Prediction of a 1990 New Madrid, Missouri, Earthquake*, USGS Circular 1082, Washington, D.C., 1993; Iben Browning et al., "Earth Tides, Volcanoes and Climatic Change," *Nature*, June 24, 1976, pp. 680–682; Mileti, 1993, pp. 52, 58–60; Sue Hubbell, "Earthquake Fever," *The New Yorker*, February 11, 1991; Pamela Sands Showalter, "Field Observations in Memphis during the New Madrid Earthquake 'Projection' of 1990: How Pseudoscience Affected a Region," Working Paper No. 71, Institute of Behavior Sciences, University of Colorado, Boulder, 1991; John Noble Wilford, "Solar System Lineup Will Bring High Tides," *New York Times*, November 27, 1990; "In Quake Zone, a Forecast Sets Off Trembles," *New York Times*, December 1, 1990; "Riding Out an Earthquake Prediction," *Washington Post*, December 2, 1990; "It's Party Time for Those Unshaken by Quake Alert," *Los Angeles Times*, December 4, 1990; "The Earthquake That Never Came," *Editor & Publisher*, December 22, 1990; Henry C. Roberts, ed., *The Complete Prophecies of Nostradamus*, Crown Publishers, 1994, pp. 305, 334; "Quaking with Anticipation," *Los Angeles Times*, May 6, 1988; "Cambodians in L.A. Area Flee, Fearing Quake," *Los Angeles Times*, May 20, 1988.

9. NORTHRIDGE (1994), PAGES 232–270

Panglossian L.A.

Charles F. Richter, "Historical Seismicity of San Fernando Earthquake Area," in National Oceanic and Atmospheric Administration (NOAA), Leonard M. Murphy, scientific coordinator, *San Fernando, California, Earthquake of February 9,*

1971, Washington, D.C., 1973; Wood, 1934, pp. 342–344; Stover, 1993, p. 124; Arthur C. Alvarez, "The Santa Barbara Earthquake of June 29, 1925," *University of California Publications in Engineering*, November 17, 1925, pp. 205–210; "Quake Proof Buildings Stressed," *Pasadena Star-News*, July 4, 1925; other newspaper stories contained in *Why Take a Chance*, California Institute of Steel Construction, undated pamphlet; Thomas H. Heaton, "Urban Earthquakes," *SRL*, September-October 1995, p. 37; Meltsner, 1979, pp. 340, 342–343, 346–348; Robert T. Hill, *Southern California Geology and Los Angeles Earthquakes*, Southern California Academy of Science, 1928; Bailey Willis, "Earthquake Risk in California," *BSSA*, vol. 14 (1924), pp. 21–23; Bailey Willis, "A Rational Basis of Earthquake Insurance," address to the National Board of Fire Underwriters, May 24, 1926; Eliot Blackwelder, "Bailey Willis," in *Biographical Memoirs*, vol. 35 (1961), National Academy of Sciences, pp. 333–342; Carl-Henry Geschwind, "1920s Prediction Reveals Some Pitfalls of Earthquake Forecasting," *EOS, Transactions, American Geophysical Union*, September 23, 1997; W. M. Davis, "The Long Beach Earthquake," *Geographical Review*, January 1934, pp. 2–11; National Board of Fire Underwriters, "Report on the Southern California Earthquake of March 10, 1933," New York City, N.Y., 1933, pp. 1–7; *Competing against Time*, 1990, pp. 117–119; Richter, 1958, pp. 388, 497–498; Carey McWilliams, "The Folklore of Earthquakes," *American Mercury*, June 1933, pp. 199–201; Carey McWilliams, *Southern California: An Island on the Land*, Peregrine Smith Books, 1973, pp. 200–204; Carey McWilliams, *California: The Great Exception*, A. A. Wyn, 1949, pp. 245–246.

The San Fernando Valley

Joint Panel on the San Fernando Earthquake, "The San Fernando Earthquake of February 9, 1971," National Academy of Sciences and the National Academy of Engineering, Washington, D.C., 1971; California Department of Water Resources, *Investigation of Failure Baldwin Hills Reservoir*, Sacramento, Calif., 1964; Jansen, 1980, pp. 120–125; "The Los Angeles Dam Story," USGS Fact Sheet 096-95, 1995; "Report of the Los Angeles County Earthquake Commission, San Fernando Earthquake, February 9," Los Angeles County Earthquake Commission, November 1971, pp. 24–25, 27–28; NOAA, vol. 2, 1973, pp. 76–77, 79, 81–82, 201–202, 235–237; Committee on Safety Criteria for Dams, "Safety of Dams," National Academy Press, Washington, D.C., 1985, pp. 39–43; Gordon W. Dukleth et al., "Seismic Safety for California Dams," *California Geology*, 1976, pp. 243–246; California State Legislature, Clarence A. Allen, "Geological and Seismological Lessons," in *The San Fernando Earthquake of February 9, 1971, and Public Policy*, Special Subcommittee of the Joint Committee on Seismic Safety, 1972, pp. 1–11, 39, 41–42; Robert L. Wiegel, *Earthquake Engineering*, Prentice-Hall, 1970, p. 188; Mihailo D. Trifunac and James N. Brune, "Complexity of Energy Release during the Imperial Valley, California, Earthquake of 1940," *BSSA*, February 1970, pp. 138, 142; William L. Ellsworth, "Earthquake History, 1769–1989," in Wallace, 1990, p. 163; interview with George Housner, Caltech, February 3, 1997; NOAA, vol. 1, 1973, pp. 255–292; Joan Didion, *Play It as It Lays*, Farrar, Straus & Giroux, 1970, pp. 15–16; Reyner Banham, *Los Angeles: The Architecture of Four Ecologies*, Penguin, 1971, pp. 89–90, 254.

The Pace Quickens

J. R. Pelton et al., "Eyewitness Account of Normal Surface Faulting," *BSSA*, June 1984, pp. 1083–1089; Robert E. Wallace, "Eyewitness Account of Surface Faulting during the Earthquake of 28 October, 1983, Borah Peak, Idaho," *BSSA*, June 1984, pp. 1091–1094; Yeats, 1997, pp. 261–269; F. Harold Weber Jr., "Whittier Narrows Earthquakes," *California Geology*, December 1987, pp. 275–281; Bill O'Callahan, "Post-Earthquake Stress: Seismic Strain on Human Psyche," *Earthquake Preparedness News (BAREPP)*, fall-winter 1987–88, p. 18; Tousson R. Toppozada, "The Landers–Big Bear Earthquake Sequence and Its Felt Effects," *California Geology*, January-February 1993, pp. 3–9; Earl W. Hart et al., "Surface Faulting Associated with the June 1992 Landers Earthquake, California," *California Geology*, January-February 1993, p. 10; Mileti, 1993, pp. 95–97; Lucile M. Jones, "Earthquake Prediction," *Proceedings of the National Academy of Science*, vol. 93 (1996), p. 3724; Ad Hoc Working Group on the Probabilities of Future Large Earthquakes in Southern California, *Future Seismic Hazards in Southern California, Phase I: Implications of the 1992 Landers Earthquake Sequence*, USGS, 1992, pp. 1–2, 4, 9, 14; "After Landers; Emotional, Physical Aftershocks Felt," *Los Angeles Times*, June 28, 1993; Mileti, 1993, pp. 94–97.

For material on the Northridge quake see John F. Hall, ed., *Northridge Earthquake of January 17, 1994, Reconnaissance Report*, vol. 1, Earthquake Engineering Research Institute, National Science Foundation, Federal Emergency Management Agency, Publication 95-03, 1995; Gore, 1995, p. 28; "Quake Swarm Provides a Jolting Surprise," *Los Angeles Times*, April 27, 1997; for more on recent death and damage totals see the *Los Angeles Times* of December 20, 1995, and March 13, 1997; interview with Thomas H. Heaton, Caltech, February 3, 1997; "After the Quake," *Newsweek*, January 31, 1994; Egill Hauksson and Lucile M. Jones, "Seismology: The Northridge Earthquake and Its Aftershocks," *Earthquakes & Volcanoes*, vol. 25, no. 1 (1941); Mary C. Woods, ed., *The Northridge, California, Earthquake of 17 January 1994*, California Division of Mines and Geology, Special Publication 116, Sacramento, Calif., 1995; "1994 Northridge Earthquake Hasn't Stopped, Hills Have Risen," Jet Propulsion Laboratory press release, NASA, December 17, 1996; Randall G.

Updike, chief scientist, USGS *Response to an Urban Earthquake, Northridge '94,*
USGS Open-File Report 96-263, Denver, Colo., 1996; Ta-liang Teng et al.,
eds., "Special Issue on the Northridge, California, Earthquake of January 17, 1994,"
BSSA, February 1996; CSSC, "Northridge Earthquake: Turning Loss to Gain,"
California Seismic Safety Commission, CSSC Report No. 95-01, 1995; House Com-
mittee on Science, Space, and Technology, *Lessons Learned from the Northridge
Earthquake,* 103rd Congress, 2nd session, March 2, 1994; Scientists of the USGS
and the Southern California Earthquake Center, "The Magnitude 6.7 Northridge,
California, Earthquake of 17 January 1994," *Science,* October 21, 1994; "Northridge,
California, Earthquake of January 17, 1994," *Earthquakes & Volcanoes,* vol. 25,
nos. 1 and 2 (1995); Mark Peterson, "The January 17, 1994, Northridge Earthquake,
California Geology, March-April 1994; "Southern Californians Cope with Earth-
quakes," USGS Fact Sheet 225-95, 1995; Dames & Moore, *The Northridge Earth-
quake January 17, 1994,* undated pamphlet; National Institute of Standards and
Technology, John L. Gross, A *Survey of Steel Moment-Resisting Frame Buildings
Affected by the 1994 Northridge Earthquake,* Building and Fire Research Laboratory,
Gaithersburg, Md., 1995; "Microstructural Characteristics of Failed Steel Moment-
Resisting Beam-Column Connections," *Earthquake Engineering Research Center
News,* University of California, Berkeley, July 1997; *Public Policy and Building Safety,*
Earthquake Engineering Research Institute, January 1996; "Doubting Thomas," *Los
Angeles Times,* May 3, 1997; interview with Cluff, July 25, 1997; CSSC, 1994,
pp. 133–138, 156–158, 167–168; Scott, "Henry J. Degenkolb," 1994, pp. 155–169;
CSSC Commissioners, "Public Safety Issues from the Northridge Earthquake of
January 17, 1994," March 1995; Mark Yashinsky et al., "The Performance of Bridge
Seismic Retrofits during the Northridge Earthquake," Caltrans Office of Earthquake
Engineering, June 1995; William P. McGowan, "Fault-line: Seismic Safety and the
Changing Political Economy of California's Transportation System," *California His-
tory,* summer 1993; Richter in NOAA, 1973, vol. 3, pp. 5–10; William T. Holmes, ed.,
"Northridge Earthquake of January 17, 1994, supplement C to vol. 2, *Earthquake
Spectra,* 1995, p. 38; "Weekly Earthquake Report for Southern California, Janu-
ary 13–19, 1994," Caltech Seismological Laboratory, January 20, 1994; "Images of
the 1994 Los Angeles Earthquake," *Los Angeles Times,* 1994; Working Group on
California Earthquake Probabilities, "Seismic Hazards in Southern California:
Probable Earthquakes, 1994–2024," *BSSA,* April 1995; R. F. Yerkes, "Geological
and Seismological Setting," in J. I. Ziony, ed., *Evaluating Earthquake Hazards in
the Los Angeles Region—an Earth-Science Perspective,* Washington, D.C., 1985,
pp. 32–33, 37.

Into the Belly of the Seismic Beast

James F. Dolan et al., "Active Tectonics, Paleoseismology and Seismic Hazards of the Hollywood Fault, Northern Los Angeles Basin, California," submitted to *BSSA*, in review February 1997; James F. Dolan et al., "Active Tectonics, Paleoseismology, and Seismic Hazards of the Hollywood Fault, Northern Los Angeles Basin, California," *GSAB*, December 1997, pp. 1595–1616; Earth Technology Corporation, "Investigations of the Hollywood Fault Zone, Segment 3, Metro Red Line," July 1993; various *Los Angeles Times* stories on tunnel mishaps, Hollywood, and the Sunset Strip, 1996–97; Woods, 1995, p. 220; interview with Stuart Warren, North Hollywood, February 10, 1997; interviews with James Monsees, chief tunnel engineer, and K. N. Murthy, project manager, February 4, 1997; Dan Eisenstein et al., "Los Angeles County Metro Rail Project: Report on Tunneling Feasibility and Performance," November 1995, pp. 8–9; "L.A.'s Oldest Residents," *Geotimes*, May 1997; "MTA Rail Construction Contributes to Scientific Knowledge," MTA press release, December 19, 1996; interview with Ronald F. Scott, Caltech, February 11, 1997; James F. Dolan et al., "Prospects for Larger or More Frequent Earthquakes in the Los Angeles Metropolitan Region," *Science*, January 13, 1995; Kerry Sieh letter to Timothy Smirnoff, February 1, 1993; Lionel Rolfe, *In Search of Literary L.A.*, California Classics Books, 1991, p. 102; "Research of the Hollywood Fault," Southern California Earthquake Center newsletter, fall 1995; James F. Dolan, "The Hollywood Fault, Revisited," Southern California Earthquake Center newsletter, spring 1997; James F. Dolan, *Field Trip Guide: Neotectonics of the Northern Los Angeles Basin*, 1997.

Universal City and the Subculture of Earthquakes

"At Universal Studios, It's Only a Movie," *New York Times*, September 29, 1996; "The Abiding Urge to Watch Things Go Ka-Boom!" *New York Times*, February 9, 1997; Howard Rabinowitz, "The End Is Near: Why Disaster Movies Make Sense (and Dollars) in the 90s," *Washington Monthly*, April 1997; Anita Loos, *Cast of Thousands*, Grosset & Dunlap, 1977, pp. 126, 133–135; Anita Loos, *San Francisco: A Screenplay*, Southern Illinois Press, 1963, pp. vii, 164–193; F. H. Weber Jr., "Earthquake," *California Geology*, June 1975, p. 135; *Earthquake*, Universal Pictures, 1974; *Escape from L.A.*, Paramount, 1996; "The Ocean Falls into L.A.," *San Francisco Chronicle*, August 9, 1996; Matthew 17:51–54, 28:2, Authorized (King James) Version; Yeats, 1997, p. 16; William Shakespeare, *Romeo and Juliet*, ed. John E. Hankins, Penguin Books, 1970, pp. 16, 42; Robert Louis Stevenson, "Aes Triplex," in *Virginibus Puerisque*, 1881, p. 112; L. Frank Baum, *Dorothy and the Wizard in Oz*, Dover Publications, 1984, pp. 20–23; John Fricke et al., *The Wizard of Oz* (pictorial history of Oz books), Warner Books, 1989; McWilliams, 1946, p. 204; F. Scott

Fitzgerald, *The Last Tycoon,* Charles Scribner's Sons, 1994, pp. x, 23; James D. Houston, *Continental Drift,* University of California Press, 1996; "Ear to the Ground," *Los Angeles Reader,* May 5, 1995.

From San Bernardino to the Gulf of California

"Northridge Quake Housing Repairs Close to Complete," *Los Angeles Times,* February 3, 1997; "Weekly Earthquake Report for Southern California," Caltech Seismology Laboratory, February 6–12, 1997; Updike, 1996, p. 7; Rob Von Zabern, Andy Jackson Air Park, February 8, 1997; "Alquist-Priolo Earthquake Fault Zoning Act," in CSSC, 1994, pp. 87–88; Lucille Wilshire Broaders, Oak Glen, February 8, 1997; Kerry E. Sieh and Jonathan C. Matti, "Earthquake Geology San Andreas Fault System Palm Springs to Palmdale," Southern California Section Association of Engineering Geologists, Los Angeles, 1992, p. 3; Woods, 1995, p. 220; "Coachella Valley Resort Targets Quake-Curious Japanese Tourists," *Riverside Press-Enterprise,* July 1, 1992; Coachella Valley Preserve and McCallum Nature Trail pamphlets, undated; Francesco V. Corona, "San Andreas Fault System," Ninth Thematic Conference on Geologic Remote Sensing, Pasadena, 1993, p. 89; Kerry E. Sieh and Patrick L. Williams, "Behavior of the Southernmost San Andreas Fault during the Past 300 Years," *JGR,* May 10, 1990; Malcolm Clark, "Map Showing Recently Active Breaks along the San Andreas Fault and Associated Faults between Salton Sea and Whitewater River–Mission Creek, California," USGS Map 1-1483, 1984; Information packet, Salton Sea State Recreation Area, undated; "Life on the Edge—of San Andreas Fault," *Desert Sun,* May 21, 1995; Philip L. Fradkin, *A River No More,* University of California Press, 1996, pp. 302–305; Bureau of Reclamation, Earl Burnette et al., *Cienega de Santa Clara Geologic and Hydrologic Comments,* Yuma, Ariz., 1993; Scott A. Zengel et al., "Cienega de Santa Clara, a Remnant Wetland in the Rio Colorado Delta (Mexico)," *Ecological Engineering,* 1995; Edward P. Glenn et al., "Cienega de Santa Clara: Endangered Wetland in the Colorado River Delta, Sonora, Mexico," *Natural Resources Journal,* fall 1992, pp. 817–824; Charles W. Jennings, "Selected Faults in Northern Baja California, Offshore, and the Adjacent Southern California Area," California Division of Mines and Geology, 1994; Ralph Lee Hopkins, "Land Torn Apart," *Earth,* February 1977; Morris A. Balderman, "The 1852 Fort Yuma Earthquake," *BSSA,* June 1978; Agnew and Sieh, 1978, appendix.

On the crisis in the field of seismology see Gordon P. Eaton, various E-mail messages, "Benchmark Notes," May 29, 1996, to March 3, 1997; "State of the Survey," Gordon P. Eaton, March 3, 1997; Dieterich, May 21, 1997; interviews with employees in USGS offices in Pasadena and Menlo Park, Calif., and Volcano, Hawaii; Richard A. Kerr, "Downsizing Squeezes Basic Research at the USGS," *Science,* June 30, 1995; Gordon P. Eaton, "What's Ahead for the USGS?" *Geotimes,* March

1996, pp. 24–26; "How Can This Be Happening?" *Washington Post Magazine,* May 19, 1996; John Maddox, "The Prevalent Distrust of Science," *Nature,* November 30, 1995; interview with William L. Condit, staff director, House Subcommittee on Energy and Mineral Resources, July 16, 1997; Thomas H. Jordan, "Is the Study of Earthquakes a Basic Science?" *SRL,* March-April 1997, pp. 259–261.

APPENDIX, PAGES 271–275

The Richter Scale Is No More

Meltsner, 1979, p. 231; McPhee, 1993, pp. 283–284; Walker, 1882, pp. 85–86; Bolt, 1993, pp. 118–123; Richter, 1958, pp. 338–365; Lucile M. Jones, *Putting Down Roots in Earthquake Country,* Southern California Earthquake Center, 1995, pp. 22–23; Gore, 1995, p. 10; USGS, National Earthquake Information Center Web Site, 1997; Rebecca Ansell and John Taber, *Caught in the Crunch: Earthquakes and Volcanoes in New Zealand,* HarperCollins New Zealand, 1996, pp. 53–54; Tousson R. Toppozada, "History of Damaging Earthquakes in Los Angeles and Surrounding Area," in Woods, 1995, p. 11; Arch C. Johnston, "An Earthquake Strength Scale for the Media and the Public," *Earthquakes & Volcanoes,* vol. 22, no. 5 (1990), pp. 214–216.

INDEX